Concepts in Molecular and Organic Electronics

MATERIALS RESEARCH SOCIETY
SYMPOSIUM PROCEEDINGS VOLUME 1154

Concepts in Molecular and Organic Electronics

Symposium held April 13–17, San Francisco, California, U.S.A.

EDITORS:

Norbert Koch
Humboldt-Universität zu Berlin
Berlin, Germany

Egbert Zojer
Technische Universität Graz
Graz, Austria

Saw-Wai Hla
Ohio University
Athens, Ohio

Xiaoyang Zhu
University of Texas at Austin
Austin, Texas

Materials Research Society
Warrendale, Pennsylvania

CAMBRIDGE
UNIVERSITY PRESS

University Printing House, Cambridge CB2 8BS, United Kingdom

One Liberty Plaza, 20th Floor, New York, NY 10006, USA

477 Williamstown Road, Port Melbourne, VIC 3207, Australia

314-321, 3rd Floor, Plot 3, Splendor Forum, Jasola District Centre, New Delhi - 110025, India

79 Anson Road, #06-04/06, Singapore 079906

Cambridge University Press is part of the University of Cambridge.

It furthers the University's mission by disseminating knowledge in the pursuit of education, learning and research at the highest international levels of excellence.

www.cambridge.org
Information on this title: www.cambridge.org/9781605111278

Materials Research Society
506 Keystone Drive, Warrendale, PA 15086
http://www.mrs.org

First published 2009
First paperback edition 2012

Single article reprints from this publication are available through
University Microfilms Inc., 300 North Zeeb Road, Ann Arbor, MI 48106

CODEN: MRSPDH

A catalogue record for this publication is available from the British Library

ISBN 978-1-605-11127-8 Hardback
ISBN 978-1-107-40833-3 Paperback

CONTENTS

PHOTOVOLTAICS

POSTER SESSION II

PROCESSING, SENSING AND MEMORY

PREFACE

This proceedings volume contains contributions presented at Symposium B, "Concepts in Molecular and Organic Electronics," held April 13–17 at the 2009 MRS Spring Meeting in San Francisco, California. Organic electronics holds the promise of applications like flexible displays, cheap to produce solar cells and printable electronic circuits. Molecular electronics, on the other hand, aims at ultimate miniaturization with monolayers or even single molecules as the active elements of electronic devices. Research in both areas has been booming over the past years. This symposium focused on stimulating the information interchange between the two communities. Its goal was to encourage scientists to adopt approaches and concepts in their own work from the "other" community. With over three hundred contributions, the resonance in the scientific community to this idea was exceedingly positive. The covered scientific topics included molecular scale electronics, where the absolute need for good statistics when evaluating experimental data was repeatedly stressed, and recent progress in this direction was demonstrated. The session dealing with interfaces as the crucial elements for understanding both macroscopic organic as well as molecular electronics clearly showed the necessity of a multi-disciplinary approach. In particular, improved theoretical approaches for interface modeling were identified to be of truly common interest to the two communities. Charge transport phenomena on several lengths scales, and new developments for different types of organic devices, namely transistors, sensors, solar cells and light emitting devices, were also reported, and the viability of the organic-based technology was highlighted. As symposium organizers, we are particularly grateful to all the participants, who made the symposium a highly stimulating scientific experience, and to all the authors who contributed papers to this proceedings.

Symposium B and this proceedings volume are dedicated to the memory of Bert de Boer. Much to everyone's regret, Bert died in late January 2009. Bert de Boer made significant contributions to the field of organic and molecular devices and was a co-organizer of the symposium. We will always remember him as a highly valued colleague and friend.

Norbert Koch
Egbert Zojer
Saw-Wai Hla
Xiaoyang Zhu

June 2009

MATERIALS RESEARCH SOCIETY SYMPOSIUM PROCEEDINGS

Volume 1153 — Amorphous and Polycrystalline Thin-Film Silicon Science and Technology — 2009, A. Flewitt, Q. Wang, J. Hou, S. Uchikoga, A. Nathan, 2009, ISBN 978-1-60511-126-1
Volume 1154 — Concepts in Molecular and Organic Electronics, N. Koch, E. Zojer, S.-W. Hla, X. Zhu, 2009, ISBN 978-1-60511-127-8
Volume 1155 — CMOS Gate-Stack Scaling — Materials, Interfaces and Reliability Implications, J. Butterbaugh, A. Demkov, R. Harris, W. Rachmady, B. Taylor, 2009, ISBN 978-1-60511-128-5
Volume 1156— Materials, Processes and Reliability for Advanced Interconnects for Micro- and Nanoelectronics — 2009, M. Gall, A. Grill, F. Iacopi, J. Koike, T. Usui, 2009, ISBN 978-1-60511-129-2
Volume 1157 — Science and Technology of Chemical Mechanical Planarization (CMP), A. Kumar, C.F. Higgs III, C.S. Korach, S. Balakumar, 2009, ISBN 978-1-60511-130-8
Volume 1158E —Packaging, Chip-Package Interactions and Solder Materials Challenges, P.A. Kohl, P.S. Ho, P. Thompson, R. Aschenbrenner, 2009, ISBN 978-1-60511-131-5
Volume 1159E —High-Throughput Synthesis and Measurement Methods for Rapid Optimization and Discovery of Advanced Materials, M.L. Green, I. Takeuchi, T. Chiang, J. Paul, 2009, ISBN 978-1-60511-132-2
Volume 1160 — Materials and Physics for Nonvolatile Memories, Y. Fujisaki, R. Waser, T. Li, C. Bonafos, 2009, ISBN 978-1-60511-133-9
Volume 1161E —Engineered Multiferroics — Magnetoelectric Interactions, Sensors and Devices, G. Srinivasan, M.I. Bichurin, S. Priya, N.X. Sun, 2009, ISBN 978-1-60511-134-6
Volume 1162E —High-Temperature Photonic Structures, V. Shklover, S.-Y. Lin, R. Biswas, E. Johnson, 2009, ISBN 978-1-60511-135-3
Volume 1163E —Materials Research for Terahertz Technology Development, C.E. Stutz, D. Ritchie, P. Schunemann, J. Deibel, 2009, ISBN 978-1-60511-136-0
Volume 1164 — Nuclear Radiation Detection Materials — 2009, D.L. Perry, A. Burger, L. Franks, K. Yasuda, M. Fiederle, 2009, ISBN 978-1-60511-137-7
Volume 1165 — Thin-Film Compound Semiconductor Photovoltaics — 2009, A. Yamada, C. Heske, M. Contreras, M. Igalson, S.J.C. Irvine, 2009, ISBN 978-1-60511-138-4
Volume 1166 — Materials and Devices for Thermal-to-Electric Energy Conversion, J. Yang, G.S. Nolas, K. Koumoto, Y. Grin, 2009, ISBN 978-1-60511-139-1
Volume 1167 — Compound Semiconductors for Energy Applications and Environmental Sustainability, F. Shahedipour-Sandvik, E.F. Schubert, L.D. Bell, V. Tilak, A.W. Bett, 2009, ISBN 978-1-60511-140-7
Volume 1168E —Three-Dimensional Architectures for Energy Generation and Storage, B. Dunn, G. Li, J.W. Long, E. Yablonovitch, 2009, ISBN 978-1-60511-141-4
Volume 1169E —Materials Science of Water Purification, Y. Cohen, 2009, ISBN 978-1-60511-142-1
Volume 1170E —Materials for Renewable Energy at the Society and Technology Nexus, R.T. Collins, 2009, ISBN 978-1-60511-143-8
Volume 1171E —Materials in Photocatalysis and Photoelectrochemistry for Environmental Applications and H_2 Generation, A. Braun, P.A. Alivisatos, E. Figgemeier, J.A. Turner, J. Ye, E.A. Chandler, 2009, ISBN 978-1-60511-144-5
Volume 1172E —Nanoscale Heat Transport — From Fundamentals to Devices, R. Venkatasubramanian, 2009, ISBN 978-1-60511-145-2
Volume 1173E —Electofluidic Materials and Applications — Micro/Biofluidics, Electowetting and Electrospinning, A. Steckl, Y. Nemirovsky, A. Singh, W.-C. Tian, 2009, ISBN 978-1-60511-146-9
Volume 1174 — Functional Metal-Oxide Nanostructures, J. Wu, W. Han, A. Janotti, H.-C. Kim, 2009, ISBN 978-1-60511-147-6

MATERIALS RESEARCH SOCIETY SYMPOSIUM PROCEEDINGS

Volume 1175E —Novel Functional Properties at Oxide-Oxide Interfaces, G. Rijnders, R. Pentcheva, J. Chakhalian, I. Bozovic, 2009, ISBN 978-1-60511-148-3

Volume 1176E —Nanocrystalline Materials as Precursors for Complex Multifunctional Structures through Chemical Transformations and Self Assembly, Y. Yin, Y. Sun, D. Talapin, H. Yang, 2009, ISBN 978-1-60511-149-0

Volume 1177E —Computational Nanoscience — How to Exploit Synergy between Predictive Simulations and Experiment, G. Galli, D. Johnson, M. Hybertsen, S. Shankar, 2009, ISBN 978-1-60511-150-6

Volume 1178E —Semiconductor Nanowires — Growth, Size-Dependent Properties and Applications, A. Javey, 2009, ISBN 978-1-60511-151-3

Volume 1179E —Material Systems and Processes for Three-Dimensional Micro- and Nanoscale Fabrication and Lithography, S.M. Kuebler, V.T. Milam, 2009, ISBN 978-1-60511-152-0

Volume 1180E —Nanoscale Functionalization and New Discoveries in Modern Superconductivity, R. Feenstra, D.C. Larbalestier, B. Maiorov, M. Putti, Y.-Y. Xie, 2009, ISBN 978-1-60511-153-7

Volume 1181 — Ion Beams and Nano-Engineering, D. Ila, P.K. Chu, N. Kishimoto, J.K.N. Lindner, J. Baglin, 2009, ISBN 978-1-60511-154-4

Volume 1182 — Materials for Nanophotonics — Plasmonics, Metamaterials and Light Localization, M. Brongersma, L. Dal Negro, J.M. Fukumoto, L. Novotny, 2009, ISBN 978-1-60511-155-1

Volume 1183 — Novel Materials and Devices for Spintronics, O.G. Heinonen, S. Sanvito, V.A. Dediu, N. Rizzo, 2009, ISBN 978-1-60511-156-8

Volume 1184 — Electron Crystallography for Materials Research and Quantitative Characterization of Nanostructured Materials, P. Moeck, S. Hovmöller, S. Nicolopoulos, S. Rouvimov, V. Petkov, M. Gateshki, P. Fraundorf, 2009, ISBN 978-1-60511-157-5

Volume 1185 — Probing Mechanics at Nanoscale Dimensions, N. Tamura, A. Minor, C. Murray, L. Friedman, 2009, ISBN 978-1-60511-158-2

Volume 1186E —Nanoscale Electromechanics and Piezoresponse Force Microcopy of Inorganic, Macromolecular and Biological Systems, S.V. Kalinin, A.N. Morozovska, N. Valanoor, W. Brownell, 2009, ISBN 978-1-60511-159-9

Volume 1187 — Structure-Property Relationships in Biomineralized and Biomimetic Composites, D. Kisailus, L. Estroff, W. Landis, P. Zavattieri, H.S. Gupta, 2009, ISBN 978-1-60511-160-5

Volume 1188 — Architectured Multifunctional Materials, Y. Brechet, J.D. Embury, P.R. Onck, 2009, ISBN 978-1-60511-161-2

Volume 1189E —Synthesis of Bioinspired Hierarchical Soft and Hybrid Materials, S. Yang, F. Meldrum, N. Kotov, C. Li, 2009, ISBN 978-1-60511-162-9

Volume 1190 — Active Polymers, K. Gall, T. Ikeda, P. Shastri, A. Lendlein, 2009, ISBN 978-1-60511-163-6

Volume 1191 — Materials and Strategies for Lab-on-a-Chip — Biological Analysis, Cell-Material Interfaces and Fluidic Assembly of Nanostructures, S. Murthy, H. Zeringue, S. Khan, V. Ugaz, 2009, ISBN 978-1-60511-164-3

Volume 1192E —Materials and Devices for Flexible and Stretchable Electronics, S. Bauer, S.P. Lacour, T. Li, T. Someya, 2009, ISBN 978-1-60511-165-0

Volume 1193 — Scientific Basis for Nuclear Waste Management XXXIII, B.E. Burakov, A.S. Aloy, 2009, ISBN 978-1-60511-166-7

Prior Materials Research Society Symposium Proceedings available by contacting Materials Research Society

Molecular Scale Electronics

Mater. Res. Soc. Symp. Proc. Vol. 1154 © 2009 Materials Research Society 1154-B04-02

A Dramatic Effect of Water on Single Molecule Conductance

Edmund Leary[1], Horst Höbenreich[1], Simon J. Higgins[1], Harm van Zalinge[1], Wolfgang Haiss[1], Richard J. Nichols[1], Christopher Finch,[2] Iain Grace[2] and Colin J. Lambert[2]
1. Department of Chemistry, University of Liverpool, L69 7ZD, U.K.
2. Department of Physics, Lancaster University, Lancaster, LA1 4YB, U.K.

ABSTRACT

Simple alkanedithiols exhibit the same molecular conductance whether measured in air, under vacuum or under liquids of different polarity[1]. Here, we show that the presence of water 'gates' the conductance of a family of oligothiophene–containing molecular wires, and that the longer the oligothiophene, the larger is the effect; for the longest example studied, the molecular conductance is over *two orders of magnitude larger* in the presence of water, an unprecedented result suggesting that ambient water is a crucial factor to be taken into account when measuring single molecule conductances (SMC), or in the design of future molecular electronic devices. Theoretical investigation of electron transport through the molecules, using the *ab initio* non-equilibrium Green's function (SMEAGOL) method[2], shows that water molecules interact with the thiophene rings, shifting the transport resonances enough to increase greatly the SMC of the longer, more conjugated examples.

INTRODUCTION

We have examined the electrical properties of molecular junctions in which a π-conjugated, often redox-active, unit is sandwiched between two thiahexyl 'spacers', with gold contacts[3-5]. Since the frontier orbitals of the alkyl groups are far from the Fermi energy of the contacts, while the frontier orbitals of the π–conjugated unit are closer to the Fermi energy, these molecules may be thought of as molecular analogs of double tunnelling barriers[6]. To examine the effect of varying the length (and hence degree of conjugation) of the π-conjugated unit, oligothiophenes were selected. Oligothiophenes, polythiophenes and their derivatives have been extensively investigated as semiconductors in organic thin film transistors and photovoltaic devices[7, 8]. We have found a remarkable effect of water on the SMC of longer oligothiophene molecules, and we report these results here.

EXPERIMENT AND THEORY

Low-coverage phases of **1–4** (as dithiolates) on clean, atomically–flat gold surfaces were prepared; X-Ray photoelectron spectra were consistent with the intact molecules 'lying down' and binding to Au through both thiol sulfurs. We employed the scanning tunnelling microscopy (STM)-based I(s) technique of Haiss *et al*[3] to determine the single molecule conductances of **1–4**. Au STM tip is brought close to the surface (without making contact), the feedback loop is switched off and the tip is retracted while the tunneling current is monitored. In experiments where a molecule bridges between tip and substrate, the current at a given retraction distance is greater than in the absence of a molecule, and characteristic current plateaux are observed; see Figure 1B.

In the theoretical calculations, the relaxed geometry of the oligothiophenes was found using the density functional code SIESTA[9]. A double-zeta plus polarization basis set, Troullier-Martins pseudopotentials to remove core electrons and the Ceperley-Alder Local Density Approximation description of the exchange correlation, were used and the atomic positions were relaxed until all force components were smaller than 0.02 eV/Å. The molecule was then extended to include the surface layers of the gold leads; a layer comprising 14 atoms was chosen and the extended region included 8 layers of gold to allow a suitable representation of charge transfer effects at a molecule-gold interface. A fixed geometry corresponding to the fully-extended, all–*trans* molecule in the junction was assumed. The location of the thiols was taken to be a top site (*i.e.* directly above a surface gold atom), although the alkane chains limit any dependence on the contact geometry.

RESULTS

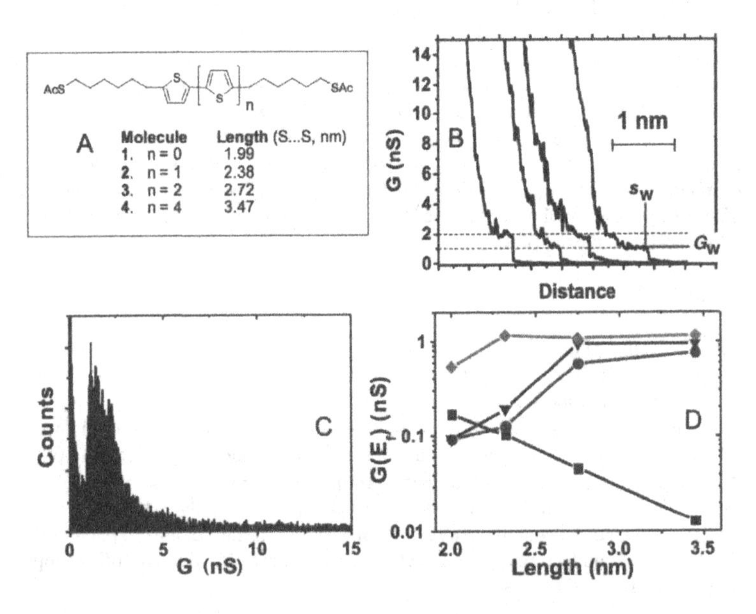

Figure 1. (A) Structures of **1-4**, and S…S distances for the fully-extended molecules, calculated using molecular mechanics. (B) Current-distance relations for several I(s) determinations on **3**. (C) Histogram of all of the characteristic conductance plateaux (G_w) observed for **2** measured in ambient air (setpoint current I_w = 4 nA, tip-sample bias 200 mV; 300 measurements, of which 72 produced plateaux as in B). (D) Experimentally-determined single molecule conductance values (*vs*. molecular length) for **1–4**, measured in ambient conditions (light blue diamonds), computed zero-bias conductance in the absence of water (black squares), and in the presence of individual water molecules located at the favoured "side" positions (red circles). The blue triangles show the computed ensemble-averaged conductance in the presence of 'side' molecules together with randomly-positioned background water molecules.

Samples of **1–4** (Figure 1A) were synthesised and fully characterised by standard chemical techniques (details to be published). Since the conduction mechanism in such junctions is superexchange, the molecular conductance of **1–4** was expected to decrease exponentially with the length of the molecule[10]. Remarkably, however, the experimental molecular conductances of **1–4** are almost length-independent (Figure 1D). It is clear from the experimental data in Figure 1D that all of the oligothiophene units are acting as tunnelling barrier indentations, because the conductances are all larger than that of $HS(CH_2)_8SH$, although **1–4** are all much longer than $HS(CH_2)_8SH$ (S…S = 1.19 nm)[11]. To understand this unusual behavior, we have carried out a detailed theoretical investigation of electron transport through these molecules, using the recently–developed *ab initio* non-equilibrium Green's function (SMEAGOL) method[12].

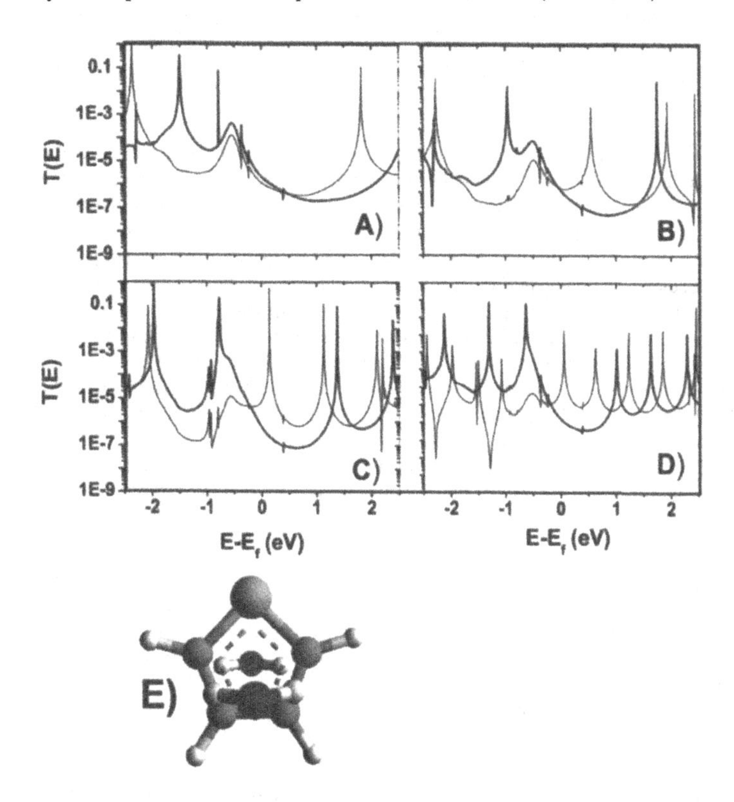

Figure 2. Zero bias transmission coefficient versus energy for **1** (A), **2** (B), **3** (C) and **4** (D). The blue curves are obtained in the absence of water and the black curves in the presence of two water molecules located at the lowest energy "side" positions as illustrated in (E).

The black curves in Figure 2(A-D) shows the electron transmission coefficient T(E) for electrons of energy E for **1–4**. These molecules exhibit the expected behavior of a decrease in the gap between the HOMO and LUMO resonances as the molecule becomes longer. Figure 1D (black squares) shows a plot of the conductance at the Fermi energy against the molecule length and shows that the conductance of **1–4** decreases exponentially with length, in spite of the decrease in HOMO-LUMO gap with increasing number of thiophene rings. The apparent discrepancy between theory and experiment was resolved by realizing that the calculations leading to Figure 1D (black squares) assume a vacuum environment, whereas the experiments were conducted in ambient air, containing water vapor, and gold electrodes under such conditions are coated with a film of water. We therefore undertook detailed *ab initio* calculations in the presence of surrounding water molecules. We examined both the relative effects of water molecules in the first solvation shell of the oligothiophene units, occupying favored local energy minima positions, and the effects of randomly-positioned water molecules beyond the first solvation shell.

First we performed molecular dynamics simulations to find the optimum location of a water molecule in the vicinity of the thiophene units. For molecule **1**, Figure 2 shows the location of the lowest energy position (in qualitative agreement with previous work[13]) and it also produces the most dramatic change in the conductance, and therefore we initially recalculated the conductance in the presence of these 'side' water molecules alone. For **1–4**, we include two 'side' water molecules per thiophene ring. Figure 2 (a-d) shows that in the presence of these 'side' molecules, the transmission resonances are significantly shifted with respect to their vacuum positions. In particular, the LUMO resonance shifts towards the Fermi energy in all molecules, thus causing an increase in the zero-bias conductances. Crucially, the longer molecules exhibit greater shifts, due to the presence of a larger number of water molecules, which in turn compensates for the reduction in conductance as a function of length. Figure 1D (red circles) shows the zero-bias conductance in the presence of water molecules in the 'side' configuration. Clearly, the exponential length dependence is removed and, in agreement with experiment, after an initial rise for small n the conductance is predicted to be almost independent of length.

We also examined the effect of a homogeneously-disordered cloud of water molecules surrounding the molecules, and we find that this alone does not remove the exponential length dependence. However, when included together with the 'side'-bound water, additional homogeneously-disordered water does have a small but significant additional effect upon the calculated conductances (Figure 1D, blue triangles). Random water also leads to a smearing of transmission resonances, so that negative differential resistance behaviour (which might be expected from a brief inspection of Figure 2C or 2D, for example) is not observed.

The remarkable results from the theoretical calculations of molecular conductance in the presence of water prompted us to perform I(s) conductance measurements on molecules **1** and **3** under water-free conditions. Figure 3 compares the conductance histograms for **3** measured after purging the STM sample chamber with dry argon for 24 h (right), and subsequently after re-admitting ambient air (left).

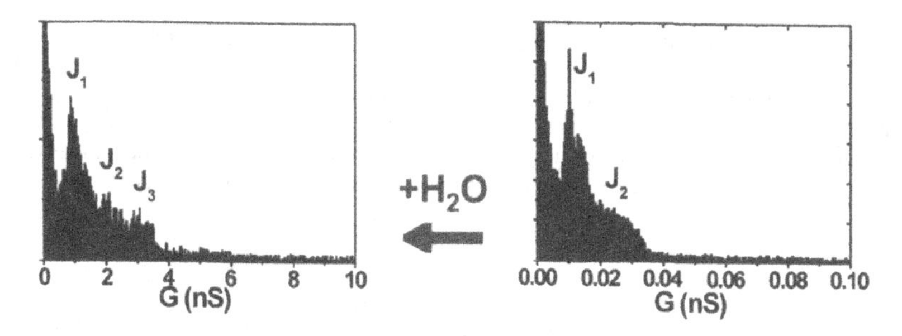

Figure 3. Histogram of the characteristic current plateaux (*I*(*w*)) observed for molecule **3** measured under: (right) dry argon, U_{tip} = +1 V, setpoint current = 7 nA; 60 I(s) scans with plateaux selected from 350 total; (left) after subsequent re-admission of ambient (wet) air to the STM chamber, same conditions, 48 I(s) scans with plateaux selected from 250 total.

The conductance of **3** determined under argon is (0.012 ± 0.006) nS *almost two orders of magnitude smaller* than the conductance in air. When ambient air was re-admitted to displace the argon in the ambient pressure STM chamber, the conductance returned to the value previously found under ambient conditions (within experimental error), (1.0 ± 0.2) nS. Crucially, this also occurred when the experiment was conducted under Ar in the presence of water vapour, excluding the possibility that atmospheric O_2 might be involved in the effect. This result is reproducible. It is important to note, however, that the conductance of **3** after argon purging for 24 h is much larger than would be predicted for an α,ω–dithiaalkane of length (2.74 nm) comparable with **3** (*i.e.* $HS(CH_2)_{20}SH$, S…S distance 2.69 nm; calc. conductance 1.37×10^{-5} nS), and that therefore, the terthiophene unit is still acting as an indentation in the tunnelling barrier. We tried to measure the conductance of junctions with molecule **4** under dry Ar, but we could not measure meaningful I(s) data; the conductance of **4** under Ar was too low for the current follower on our STM even with U_{tip} = 1.0 V. This puts an upper limit of the conductance of **4** in the absence of water of 0.006 nS, over two orders of magnitude smaller than in the presence of water. In contrast, and again in agreement with the theoretical calculations, the molecular conductance of **1** did not change significantly under argon compared with ambient air.

DISCUSSION

The observation that water, by weakly interacting with the π–system of a molecule, can drastically alter the molecular conductance of the gold|molecule|gold junction, has wider implications in other important electron transfer processes. Water is known significantly to affect rates of electron transfer (ET) reactions in biological systems by mediating ET coupling pathways and changing activation free energies[14]. Unusual 'structured' water molecules near redox cofactors have been found to accelerate ET kinetics when the cofactors are close together; such water appears to show anomalously weak distance decay for electron tunneling[15]. In this instance, it is believed that water is involved in the electron tunneling pathway, whereas in **3** and **4** it "gates" the electron tunnelling pathway through the molecule. It has been suggested that water could interact strongly with thiol–gold contacts[16], leading to a length-independent *lowering* of the conductance of gold|molecule|gold junctions, for simple alkanethiols and alkanedithiols. Our results represent the first examples of a length-dependent water-induced "gating" of electron transport through a synthetic molecular wire. They will have implications for molecular electronics, and it may also provide a new paradigm for the construction of single molecule sensors. It also emphasizes that solvent effects must be taken into account in laboratory studies of single molecule electronics.

REFERENCES

1. X. L. Li, J. He, J. Hihath, B. Q. Xu, S. M. Lindsay and N. J. Tao, J. Am. Chem. Soc. **128**, 2135-2141 (2006).
2. A. R. Rocha, V. M. Garcia-Suarez, S. Bailey, C. Lambert, J. Ferrer and S. Sanvito, Phys. Rev. B **73**, 085414 (2006).
3. W. Haiss, H. van Zalinge, S. J. Higgins, D. Bethell, H. Höbenreich, D. J. Schiffrin and R. J. Nichols, *J. Am. Chem. Soc.* **125**, 15294-15295 (2003).
4. E. Leary, S. J. Higgins, H. van Zalinge, W. Haiss, R. J. Nichols, J. O. Jeppesen, S. Nygaard and J. Ulstrup, J. Am. Chem. Soc. **130**, 12204-12205 (2008).
5. E. Leary, S. J. Higgins, H. van Zalinge, W. Haiss and R. J. Nichols, Chem. Comm. 3939-3942 (2007).
6. L. L. Chang, L. Esaki and R. Tsu, *Appl. Phys. Lett.*, **24**, 593-595 (1974).
7. T. A. Skotheim and J. R. Reynolds, eds., Handbook of Conducting Polymers Third Edition, Taylor and Francis Group, Boca Raton, 2007.
8. I. McCulloch, M. Heeney, C. Bailey, K. Genevicius, I. Macdonald, M. Shkunov, D. Sparrowe, S. Tierney, R. Wagner, W. M. Zhang, M. L. Chabinyc, R. J. Kline, M. D. McGehee and M. F. Toney, Nature Materials **5**, 328 (2006).
9. J. M. Soler, E. Artacho, J. D. Gale, A. Garcia, J. Junquera, P. Ordejon and D. Sanchez-Portal, *J. Phys-Condens. Mat.* **14**, 2745-2779 (2002).
10. A. M. Kuznetsov and J. Ulstrup, Electron Transfer in Chemistry and Biology: An Introduction to the Theory, John Wiley, Chichester, U.K., 1999.
11. W. Haiss, S. Martín, E. Leary, H. van Zalinge, S. J. Higgins, L. Bouffier and R. J. Nichols, *J. Phys. Chem. C*, 2009, **113**, 5823-5833.
12. A. R. Rocha, V. M. Garcia-Suarez, S. W. Bailey, C. J. Lambert, J. Ferrer and S. Sanvito, Nature Materials **4**, 335-339 (2005).
13. S. C. Meng, J. Ma and Y. S. Jiang, J. Phys. Chem. B **111**, 4128-4136 (2007).
14. H. B. Gray and J. R. Winckler, Q. Rev. Biophys. **36**, 341 (2003).
15. J. P. Lin, I. A. Balabin and D. N. Beratan, Science **310**, 131 (2005).
16. D. P. Long, J. L. Lazorcik, B. A. Mantooth, M. H. Moore, M. A. Ratner, A. Troisi, Y. Yao, J. W. Ciszek, J. M. Tour and R. Shashidhar, Nature Materials **5**, 901-908 (2006).

Poster Session I

Mater. Res. Soc. Symp. Proc. Vol. 1154 © 2009 Materials Research Society 1154-B05-07

Layer Cross-Fading at Organic/Organic Interfaces in OVPD-Processed Red Phosphorescent Organic Light Emitting Diodes as a New Concept to Increase Current and Luminous Efficacy

Florian Lindla[1*], Manuel Boesing[1], Christoph Zimmermann[1], Frank Jessen[1], Philipp van Gemmern[2], Dietrich Bertram[2], Dietmar Keiper[3], Nico Meyer[3], Michael Heuken[1,3], Holger Kalisch[1], Rolf H. Jansen[1]

[1]Chair of Electromagnetic Theory, RWTH Aachen University, Kackertstraße 15-17, 52072 Aachen, Germany
[2] Philips Technologie GmbH, Philipsstraße 8, 52016 Aachen, Germany
[3]AIXTRON AG, Kaiserstraße 100, 52134 Herzogenrath, Germany
[*]Corresponding author: florian.lindla@rwth-aachen.de

ABSTRACT

The current and luminous efficacy of a red phosphorescent organic light emitting diode (OLED) with sharp interfaces between each of the organic layers can be increased from 18.8 cd/A and 14.1 lm/W (at 1,000 cd/m^2) to 36.5 cd/A (+94%, 18% EQE) and 33.7 lm/W (+139%) by the introduction of a layer cross-fading zone at the hole transport layer (HTL) to emission layer (EL) interface. Layer cross-fading describes a procedure of linearly decreasing the fraction in growth rate of an organic layer during deposition over a certain thickness while simultaneously increasing the fraction in growth rate of the following layer. For OLED processing and layer cross-fading organic vapor phase deposition (OVPD) is used. The typical observation of a roll-off in current efficacy of phosphorescent OLED to higher luminance can be reduced significantly. An interpenetrating network of a prevailing hole and a prevailing electron conducting material is created in the cross-fading zone. This broadens the recombination zone and furthermore lowers the driving voltage. The concept of layer cross-fading to increase the efficacies is suggested to be useful in multi-colored OLED stacks as well.

INTRODUCTION

OLED have the potential to play a dominant role in solid state lighting. One of the most important tasks to achieve higher efficiencies in OLED is to investigate and understand the injection at the electrodes and the electronic interaction at the interfaces between the different organic layers and materials.

An optimized structure, in which the recombination zone is not only located at the interfaces, but significantly broadened has to be found to increase OLED efficacy [1]. This can be done using a mixture of electron and hole conducting host materials or an unipolar conducting host material in combination with an emitter, which acts as an hole or electron conductor [2]. As well as an increased lifetime, the typical roll-off of phosphorescent OLED due to triplet-triplet annihilation processes is reduced by a broadened recombination zone [3]. In this work, the effect of layer cross-fading on current and luminous efficacy, roll-off and lifetime is studied and compared with mixed host interlayers of constant ratio. IV measurements of mixtures of the different matrix materials are performed to gather an insight in the charge carrier mobility change at different mixture ratios.

Cross-faded host material compositions with different and complex profiles can easily be realized by using OVPD for OLED processing. A detailed knowledge about the influence of main process parameters on the morphology of organic layers [4] is necessary to reproducibly control and investigate cross-fading zones.

EXPERIMENTAL

All OLED are processed by OVPD, which allows to control all important growth parameters (substrate temperature, deposition chamber pressure, carrier gas flows) individually and guarantees stable growth rates over a long period of time.

The base structure of the investigated red phosphorescent OLED can be seen in Fig. 1.

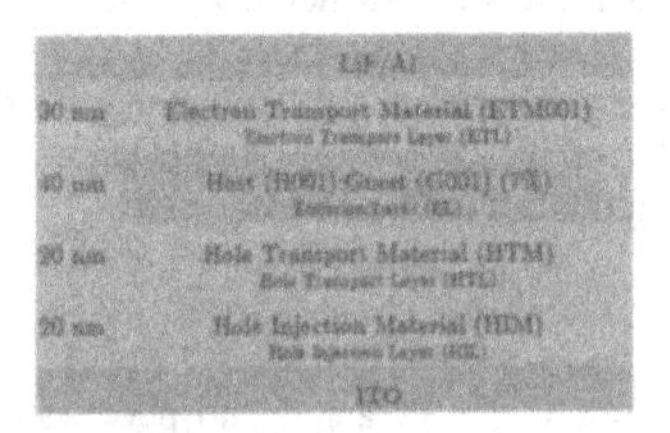

Fig. 1: Red phosphorescent OLED base structure

ITO-covered glass substrates are pre-cleaned and the ITO is oxygen plasma activated. Except when stated otherwise, all OLED are processed at a substrate temperature of 75°C and a deposition chamber pressure of 0.9 mbar.

The possibility to control the carrier gas flows and thereby the growth rate of each organic material individually is used to decrease the fraction in growth rate of a depositing organic layer over a certain thickness while increasing the fraction in growth rate of the following layer. Sharp interfaces in the reference device are cross-faded with this approach. An OLED layer structure with a cross-fading zone from HTL to EL comprising host and guest at a constant ratio can be seen in Fig. 2 (left).

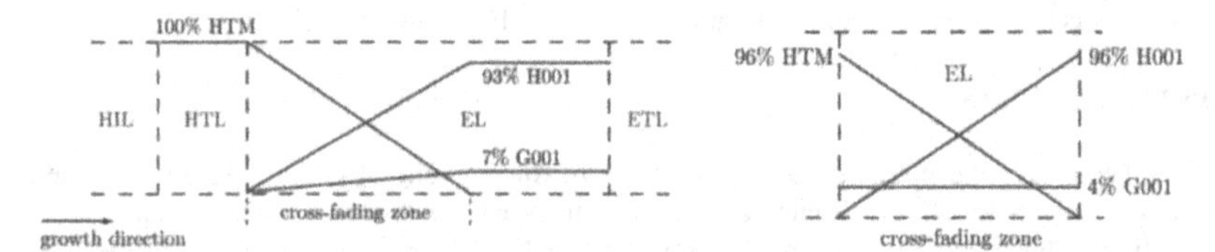

Fig. 2: Layer structure with cross-fading zone (left), cross-fading zone with constant dopant ratio (right)

It is also possible to dope the whole cross-fading zone precisely with a constant fraction of emitter molecules. The total growth rate of the cross-faded host materials can always be kept constant. Fig. 2 (right) shows this different approach. The cross-fading zone spans over the complete EL.

Material curves in Fig. 2 symbolize the fraction of each material in the organic layer at a certain position. This fraction depends on the growth rates which are connected to the source carrier gas flows changed during deposition. Even if the carrier gas flows in the used regimes are nearly linearly connected to the growth rates, this is normally not the case and the shown curves especially at very low growth rates are idealized. The exact growth rates can be looked up in the growth rate versus source carrier gas flow calibration data.

To predict the location of the recombination zone in the cross-fading zone, hole- and electron-only devices according to Fig. 3 are processed and mobilities are calculated from IV measurements.

	Al		LiF/Al
10 nm	HTM	10 nm	ETM001
50 nm	HTM:H001/x%	20 nm	H001:HTM/x%
10 nm	HTM	10 nm	H001
	ITO		ITO

Fig. 3: Layer structure of hole-only (left) and electron-only (right) devices

The Fermi-level of a non-doped aluminum cathode of 4.3 eV [5] together with the lowest unoccupied molecular orbital (LUMO) of HTM of 2.4 eV [6] will block electrons in the hole-only devices (Fig. 3 left), while the highest occupied molecular orbital (HOMO) of H001 of 5.9 eV [6] will prevent the injection of holes from the ITO (Fig. 3 right) in case of the electron-only devices.

Photometric measurements are performed using photodiodes calibrated by a Minolta Luminance Meter LS-110, while voltages and currents are supplied and measured by a Keithley 2400 Source Meter. Except when stated otherwise, all efficacy figures are measured at a luminance of 1,000 cd/m^2.

RESULTS AND DISCUSSION

For comparison, a red phosphorescent reference OLED with sharp interfaces is processed layer-by-layer according to Fig. 1. The driving voltage is 4.18 V at a current density of 5.32 mA/cm^2. This leads to a current efficacy of 18.8 cd/A and a luminous efficacy of 14.1 lm/W. All OLED investigated have the same color point of CIE 0.62/0.38.

The introduction of a cross-fading zone of 10 nm thickness from either the HIL to HTL or EL to ETL has no impact on the efficacies.

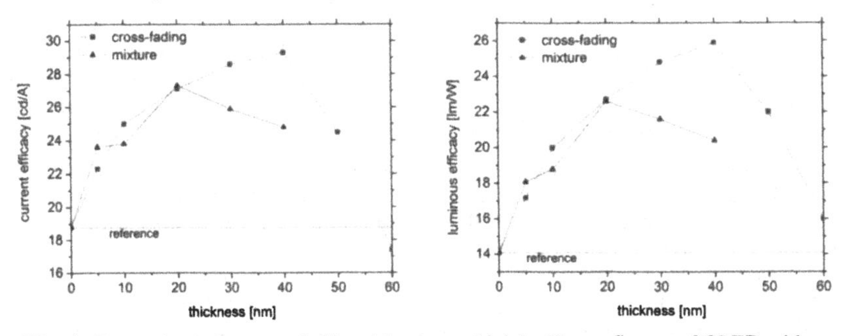

Fig. 4: Comparison of current (left) and luminous (right) efficacy figures of OLED with a cross-fading zone or mixed interlayer between HTL and EL

Contrary behavior can be observed at the HTL to EL interface. In Fig. 4 (left) the current efficacy of OLED with a cross-fading zone at the HTL to EL interface of a thickness between 5 nm and 60 nm is compared with OLED in which the cross-fading zone is replaced by a mixed interlayer of the same thickness and a constant ratio of HTM:H001 of 1:1.

Up to a cross-fading zone thickness of 20 nm, the mixed interlayer OLED show comparable efficacies, whereas from then on, a rising cross-fading zone still increases the efficacies while the mixed interlayer efficacies begin to decrease again. The maximum in current efficacy of 29.3 cd/A and a luminous efficacy of 25.9 lm/W can be measured at a cross-fading zone thickness of 40 nm. The voltage drops from 4.07 V at a cross-fading zone thickness of 5 nm to 3.56 V at 40 nm and 3.43 V at 60 nm, while the current density drops from 4.48 mA/cm^2 to 3.4 mA/cm^2 at 40 nm and increases again to 5.76 mA/cm^2 at 60 nm.

To evaluate the mobilities of holes and electrons in mixtures of HTM and H001 and thereby in certain parts of a cross-fading zone, hole- and electron-only devices according to Fig. 3 are processed.

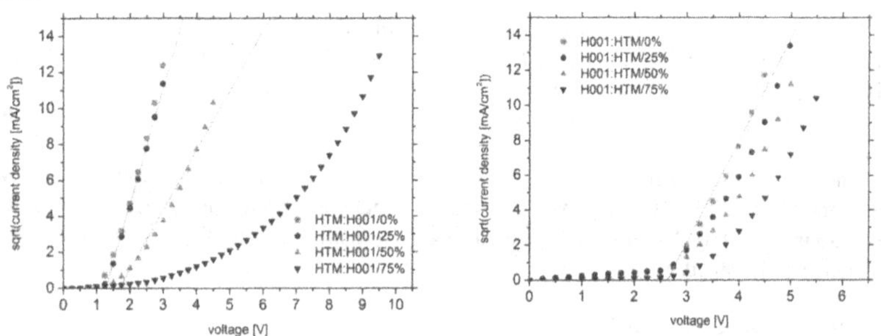

Fig. 5: IV curves of HTM:H001 hole-only devices (left) and H001:HTM electron-only devices (right)

In Fig. 5 (left), the IV curves of HTM:H001 hole-only devices with different H001 concentrations are shown. In case of a space charge limited current, current density over voltage can be described by Mott-Gurney equation [7]:

$$J = \frac{9}{8}\varepsilon\varepsilon_0\mu\frac{(V - V_{bi})^2}{d^3} \tag{1.1}$$

IV curves measured at pure HTM layers or mixed with 25% of H001 show a linear behavior indicating a space charge limited current. This is no longer the case above 50% H001 concentration. The hole mobility at 25% H001 concentration can be calculated according to (1.1) to $\mu_h = 2\cdot10^{-5}$ cm^2/Vs decreasing to $\mu_h = 5\cdot10^{-6}$ cm^2/Vs at 50%. In all calculations, $\varepsilon = 3$ is used [7]. Above 50% H001 concentration, the Mott-Gurney law can no longer be applied, which is evidenced by non-linear curves in Fig. 5 (left). It is nevertheless obvious that the current density above a concentration of 75% drops to negligible values at voltages used to drive the cross-fading devices. IV curves of electron-only devices can be seen in Fig. 5 (right). None of the devices shows a completely linear behavior, which would indicate purely space charge limited current. This might be the result of the complex device structure including ETM001 as ETL. If nevertheless the mobility of pure H001 is calculated, it results in $\mu_e = 9\cdot10^{-7}$ cm^2/Vs which is less compared to the hole mobility of HTM in a HTM:H001/50% mixture. As a result, the main part of the recombination will be located in the middle to cathode side of the cross-fading zone of devices shown in Fig. 2.

In the cross-fading zone, an interpenetrating network of the prevailing hole conducting material HTM and the prevailing electron conducting material H001 is created. Exciton formation on G001 emitter molecules becomes probable over a wider range. As result, the current efficacy increases.

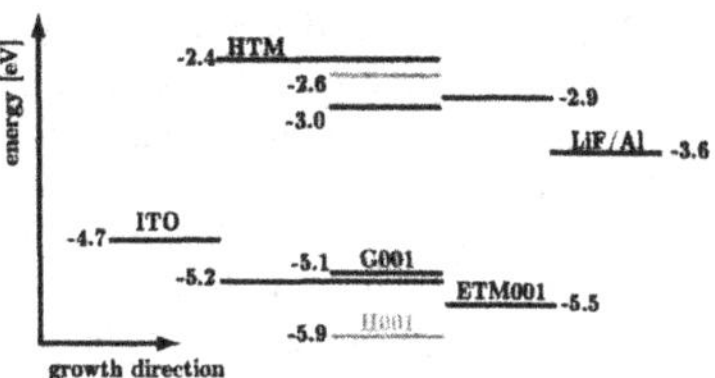

Fig. 6: Energy levels [6] of involved organic materials

The HOMO and LUMO levels of the involved organic materials are shown in Fig. 6. HTM with a high HOMO energy level and H001 with a low LUMO energy level form an ideal hole and electron conducting host for the G001 emitter and thereby reduce the voltage drop in the device from 4.18 V to around 3.56 V. This is roughly the HTM HOMO to H001 HOMO difference.

A broadened recombination zone should have a less steep decrease in current efficacy as commonly seen in phosphorescent devices due to triplet-triplet annihilation processes. Fig. 7 (left) shows the current efficacy versus luminance for the basic layer-by-layer processed OLED, one with a 40 nm cross-fading zone and one with a 40 nm mixed interlayer.

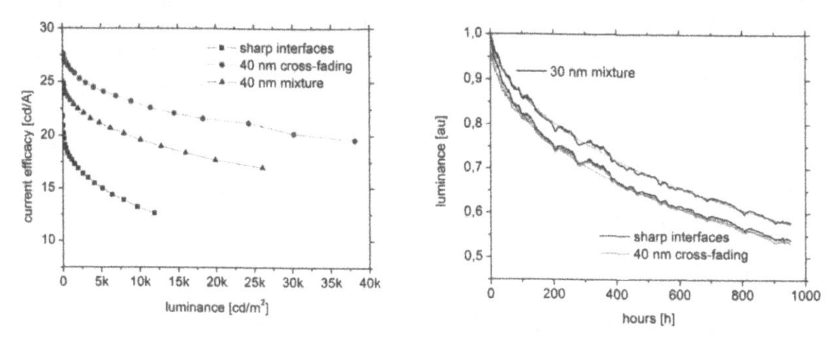

Fig. 7: Comparison of current efficacy roll-off (left) and lifetime measurements (right) for sharp interfaces, a cross-fading zone and a mixed interlayer

The roll-off in current efficacy of the OLED with sharp interfaces is significantly steeper compared to the one with a cross-fading zone. The current efficacy at a luminance increase from 100 to 10,000 cd/m^2 decreases with sharp interfaces to 65% and with a cross-fading or mixture zone to 82-85%. This supports the assumption of a broadened recombination zone spread over a significant part of the cross-fading zone between HTL und EL.

If the luminance at a constant and equal current for sharp interfaces, a 40 nm cross-fading and a 30 nm mixture zone is measured over time, a lifetime of 1,200 hours for the sharp interfaces as well as the cross-fading zone and 1,400 hours for the mixture zone can be extrapolated in Fig. 7 (right). The lifetime is here defined as a decrease in luminance to 50% of the initial value. The total electrical stress at a constant current should be the same for all devices. Surprisingly, the lifetime of the cross-faded device has not increased compared to the layer-by-layer processed one. A different lifetime test, which compared a sharp interface device with a 10 nm cross-fading zone and a 10 nm mixture zone results in a cross-fading device lifetime between that of the sharp interface and the mixture zone device. The reason for the unimproved lifetime of the 40 nm cross-fading zone device remains unclear.

Besides the cross-fading profile in Fig. 2 (left), several different profiles are tested. The profile of the dopant G001 is less important, because the emitter is not necessary for hole conduction. Therefore, a constant dopant ratio is used. With a cross-fading zone profile as shown in Fig. 2 (right) and a constant dopant concentration of 4%, finally an EQE of 18% at a current efficacy of 36.5 cd/A is reached. The cross-fading EL is sandwiched between 40 nm HTM as HTL and 20 nm H001 plus 20 nm ETM001 as ETL.

SUMMARY AND CONCLUSIONS

The introduction of a cross-fading zone at the HTL to EL interface can significantly increase current and luminous efficacy of red phosphorescent OLED. Compared to a layer-by-layer processed structure with sharp interfaces, 29.3 cd/A (+56%) and 25.9 lm/W (+84%) can be measured. HTM and H001 form an energetically-ideal suited mixed matrix for the red emitter G001, which lowers the driving voltage. By optimizing and spreading the cross-fading zone over the whole EL, the OLED efficacy could be further increased to 36.5 cd/A (18% EQE) and 33.7 lm/W. Achieving a broadened recombination zone by cross-fading a prevailing hole into a prevailing electron conducting material should always be advantageous in case of appropriate matrix materials. The typically observed roll-off of phosphorescent OLED can be reduced.

In future, the layer cross-fading concept can be used to form an EL basis for two different kinds of phosphorescent emitters with the possibility of an easy color tuning.

ACKNOWLEDGMENT

Financial support by the German Bundesministerium für Bildung und Forschung (OPAL project grant number 13N8669) is gratefully acknowledged.

REFERENCES

[1] Kondakova, M.E., et al., J. of Appl. Phys. (2008) **104**, 094501
[2] Adachi, C., et al., Organic Electronics (2001) **2**, 37
[3] Kim, S.H., et al., Appl. Phys. Let. (2008) **92**, 02351
[4] Boesing, M, et al., Mat. Res. Soc. Symp. Proc. (2008) **1115**, in press
[5] Stöcker, Taschenbuch der Physik, Verlag Harri Deutsch (2000)
[6] Merck, Technical Data Sheet, not published
[7] Brüttig, W., Physics of Organic Semiconductors, Wiley-VCH (2005)

Mater. Res. Soc. Symp. Proc. Vol. 1154 © 2009 Materials Research Society 1154-B05-11

Correlation Between On/Off Ratio and Electron Traps in Hole-Only Carbon-Nanotube-Enabled Vertical Field Effect Transistors

Mitchell A. McCarthy[1], Bo Liu[2] and Andrew G. Rinzler[2]

[1] Department of Materials Science and Engineering, University of Florida, Gainesville, FL 32611

[2] Department of Physics, University of Florida, Gainesville, FL 32611

ABSTRACT

Single wall carbon nanotube enabled vertical field effect transistors (VFETs) are studied and the dependence of the on/off ratio on the relative number of electron traps is investigated. Current versus voltage measurements on several VFETs with varying interfacial trap densities in the vicinity of the nanotube network/polymer active layer junction are taken. It is found that the on/off ratio of the VFET changes from 1600 to 20 for typical operational currents as the onset gate voltage in the off-to-on transfer curve shifts from 94 V to 72 V. Such a strong dependence on trapped charge motivates future work to uncover the mechanism of charge trapping.

INTRODUCTION

Efforts to commercialize active matrix organic light emitting diode (AMOLED) displays are increasing and motivated by the promise of mechanical flexibility and increased viewing angle and response time [1-3]. Organic light emitting transistors have been demonstrated [4-6] and promise to reduce the driving voltage and simplify the circuitry in all-organic AMOLED displays. Among the promising candidates for organic light emitting transistors is the carbon nanotube enabled vertical organic light emitting transistor (VOLET). Initial nanotube enabled VOLETs showed large hysteresis with a two orders of magnitude on/off ratio [4]. Hysteresis in the transfer curve results from charge trapping in the vicinity of the hole-injection electrode/active layer junction of the device. This trapped charge in-turn screens the electric field created by the gate voltage reducing the on/off ratio. In the present study it is demonstrated that electron traps play a more significant role in reducing the on/off ratio than hole traps. Electron traps are relevant when the gate voltage is positive and act to shift the threshold voltage of the transistor when the gate voltage is swept from positive to negative (off-to-on sweep direction for a p-channel transistor). Reducing the number of trapped electrons in this unconventional transistor architecture, reduces the off-to-on threshold voltage and increases the on/off ratio.

In previous studies [4], vertical field effect transistors (VFETs) together with VOLETs were studied. The VFET studied shares the same hole-injection electrode (a dilute nanotube network) and polymer active layer as the VOLET. The difference between the two is that the VOLET has a low work function electron injecting top contact above an electroluminescent polymer which makes it a dual carrier device; the VFET is a single carrier (hole-only) device because of the degree of alignment of the work function of the nanotube and the highest occupied molecular orbital (HOMO) level of the polymer active layer, enabling hole-injection, and the large misalignment between the polymer lowest unoccupied molecular orbital (LUMO) and the gold top contact, creating a large barrier preventing electron injection. The transconductance mechanism is the same for the VFET and the VOLET because both involve the nanotube/polymer active layer interface. Modulation of the gate electric field moves the Fermi level inside the polymer active layer and the nanotube source electrode modulating the barrier height and width for hole-injection from the nanotube electrode into the polymer layer, resulting in transconductance [4]. To study trapping at the nanotube/polymer layer junction we built hole-only VFETs instead of VOLETs—for removal of unnecessary device complexity. Hole-only VFETs are sufficient to study hysteresis and on/off ratio effects because the dominant gating mechanism occurs in the region of the interface between the hole injection electrode and the active layer. Therefore, knowledge gained from hole-only VFET devices can be directly applied to VOLETs. Additionally, it was verified in a previous study [4], that the general shape of the hysteresis in the transfer curve was the same in a VFET and VOLET.

Difficult to control minor variability in device processing steps revealed behavior in the devices that strongly imply a modification of the interfacial trap density in the region of the nanotube/polymer junction. Several VFETs were fabricated, one of which received a spin-on-glass (SOG) coating on the nanotube network which was being investigated for other purposes (not discussed here), and then was subsequently removed in warm KOH solution. Transconductance measurements are taken on each VFET and a correlation between the off-to-on threshold voltage in the transfer curves and the on/off ratio at a typical operation current is reported. Atomic force microscopy (AFM) is used to verify the morphology of the nanotube network as well as the polymer active layer thickness.

EXPERIMENTAL DETAILS

Highly doped p-type Si (<.005 ohm-cm) substrates with a 200 nm LPCVD silicon nitride layer were cleaned and chromium (7 nm) and gold (45 nm) bottom contacts were thermally evaporated onto the substrate to serve as contacts to the nanotube network. Single wall carbon nanotubes were pulsed laser vaporization grown and purified as described elsewhere [7]. Nanotube material was vacuum filtered onto a mixed cellulose ester membrane then rinsed in DI water and dried and transferred to the substrate by the methods in [8] to form a dilute network which is well above the percolation threshold (AFM image shown in Figure 1B) with a sheet resistance of ~5 kohms (as determined by the 4 terminal Van der Pauw technique). Poly((9,9-dioctylfluorenyl-2,7-diyl)-alt-co-(9-hexyl-3,6-carbazole)) (PF-9HK) from American Dye Sources at 2.5-4.0 wt% in toluene was spin coated to a thickness of ~200-350 nm in an Ar glovebox with water and oxygen levels less than .1 ppm. The polymer coated nanotube substrates were transferred into an evaporator integrated into an Ar filled glovebox. A top contact consisting of 20 nm of gold was deposited through a shadow mask defining discrete pixels with an area of 3.5×10^{-4} cm^2 to allow the measurement of multiple device regions (Figure 1A shows the VFET schematic). The SOG pretreatment (SOG on/off) step used KASIL 1 (potassium silicate) from PQ Corporation diluted in deionized (DI) water (>18.1 Mohm-cm) and filtered through a .2 μm PTFE pore size membrane and spin coated onto the substrate at ~1000-6000 rpm. The SOG coated substrate was cured at 80° C for 30 min then at 170° C for 30 min. KOH solution (5 wt. % in DI water) at 48° C was used to remove the SOG by submersion of the sample for 3 min. VFETs were measured with a homebuilt transconductance measurement system with source-drain and gate leakage currents read by Keithley model 414s and 485 picoammeters and signals provided and read by a National Instruments PCI-MIO-16XE-10 multifunction card controlled by a program written in LabVIEW.

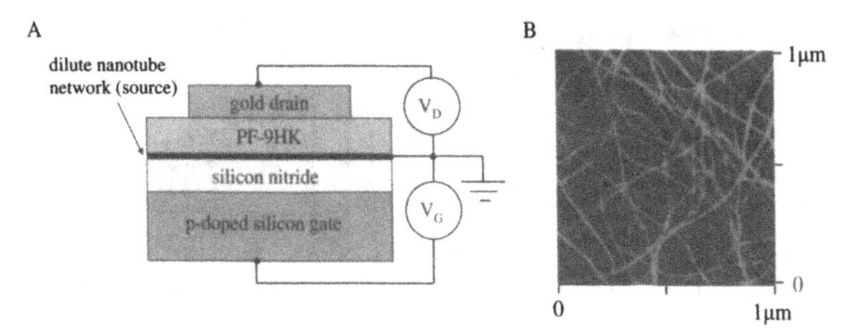

Figure 1. VFET schematic (A) and an AFM image of the dilute nanotube network used as the source electrode (B) (~5 kohms sheet resistance).

DISCUSSION

Assessments of the sign and relative amount of trapped charge can be determined from hysteresis data. In the transfer data in Figure 2A, when the gate voltage (V_G) is swept from 100 V to -100 V (off-to-on sweep direction), the drain current (I_D) quickly turns on and then reaches a maximum and flattens out. When V_G is negative, holes becomes trapped in the vicinity of the nanotube/PF-9HK interface. Upon reversal (on-to-off), the trapped holes cause the drain current to quickly turn off, as they are creating their own electric field that acts to increase the barrier height and width for hole-injection from the nanotube network into the PF-9HK. When V_G becomes positive, electrons are trapped and screen the gate electric field thereby preventing the drain current from turning off further. The relative number of trapped electrons between devices can be inferred by the threshold voltage in the off-to-on curves in Figure 2A. The larger V_G is when the drain current begins to increase the more electron traps exist at the interface. With more trapped electrons the on/off ratio of the device is reduced as shown in Figure 2B. The on/off ratio is calculated from output curves (like the ones shown in Figure 2D) where the drain voltage (V_D) is swept from 0 to -18 V in 20 steps beginning with V_G= 100V and in each subsequent curve V_G was decremented by 25 V until it reached -100 V, where the on/off ratio = $I_{on(VG=-100\ V)}/I_{off(VG=100\ V)}$. Figure 2C combines the relevant data of Figure 2A and Figure 2B; it plots the on/off ratio at an I_D = 5 mA/cm^2 (in the on-state) from Figure 2B vs. the V_G when I_D in the off-to-on transfer curve surpasses .25 mA/cm^2 from Figure 2A. An I_D of .25 mA/cm^2 is chosen as a rough indicator of when light would begin to be visible in a light emitting material with current efficiency in the range of 1-5 cd/A in the VOLET. The variation in the number of trapped electrons causes a nearly 2 order of magnitude change in the on/off ratio of the VFET. A variation in the position of the on-to-off transfer curves suggests a change in the number of hole traps, however, the change does not follow a consistent trend.

The SOG on/off device has the highest on/off ratio as well as a transfer curve with an additional inflection point (at V_G ~ -62 V) and an obvious maximum at V_G ~ -25 V. The

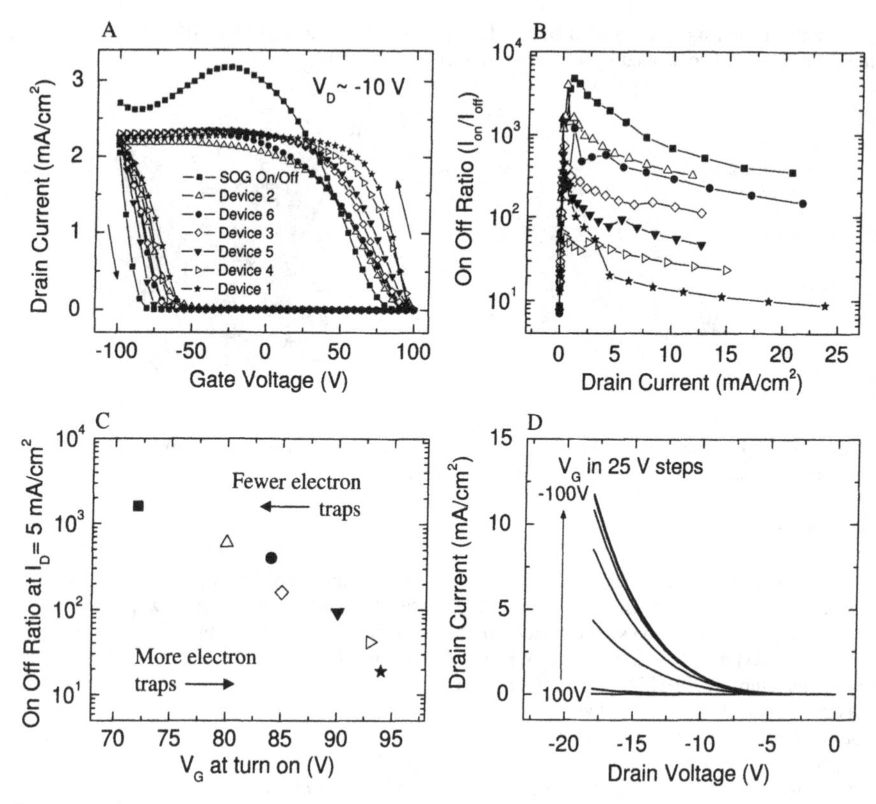

Figure 2. Transfer curves (A) and on/off ratio curves (B) for the devices. The on/off ratio curves in (B) are plotted as a function of I_D in the on-state of the device (V_G = -100 V). Plot of the on/off ratio (C) at 5 mA/cm^2 drain current (in the on-state) vs. the V_G when the current in the off-to-on curve in (A) goes above .25 mA/cm^2. Output curves (D) for device 2 represent typical behavior of the VFET.

maximum is attributed to a larger trapping relaxation time resulting most likely from left over potassium silicate on the nanotube network, and/or increased trapping of holes at negative values of V_G causing screening and reducing I_D.

CONCLUSIONS

The nanotube enabled VOLET is a newly realized type of light emitting transistor that shows promise in all-organic AMOLED displays. The nanotube enabled VFET suffices to study the transconductance mechanisms governed by the nanotube/polymer interface; and knowledge gained from the VFET can be directly applied to the VOLET. The nanotube enabled VFET operates similarly to a p-channel FET and has large hysteresis. The off current is very sensitive to the number of electron traps causing the on/off ratio to vary from 20 to 1600. With the benefits of reducing the number of electron traps made apparent, future efforts will focus on means to reduce them.

ACKNOWLEDGEMENTS

This material is based upon work supported by the National Science Foundation under Grant No. 0824157.

REFERENCES

[1] T. Ishibashi, J. Yamada, T. Hirano, Y. Iwase, Y. Sato, R. Nakagawa, M. Sekiya, T. Sasaoka, and T. Urabe, "Active matrix organic light emitting diode display based on "Super Top Emission" technology," *Japanese Journal of Applied Physics Part 1-Regular Papers Brief Communications & Review Papers,* vol. 45, pp. 4392-4395, May 2006.

[2] M. Mizukami, N. Hirohata, T. Iseki, K. Ohtawara, T. Tada, S. Yagyu, T. Abe, T. Suzuki, Y. Fujisaki, Y. Inoue, S. Tokito, and T. Kurita, "Flexible AM OLED panel driven by bottom-contact OTFTs," *Ieee Electron Device Letters,* vol. 27, pp. 249-251, Apr 2006.

[3] S. Ohta, T. Chuman, S. Miyaguchi, H. Satoh, T. Tanabe, Y. Okuda, and M. Tsuchida, "Active matrix driving organic light-emitting diode panel using organic thin-film transistors," *Japanese Journal of Applied Physics Part 1-Regular Papers Short Notes & Review Papers,* vol. 44, pp. 3678-3681, Jun 2005.

[4] B. Liu, M. A. McCarthy, Y. Yoon, D. Y. Kim, Z. Wu, F. So, P. H. Holloway, J. R. Reynolds, J. Guo, and A. G. Rinzler, "Carbon-Nanotube-Enabled Vertical Field Effect and Light-Emitting Transistors," *Advanced Materials,* vol. 20, 2008.

[5] K. Nakamura, T. Hata, A. Yoshizawa, K. Obata, H. Endo, and K. Kudo, "Metal-insulator-semiconductor-type organic light-emitting transistor on plastic substrate," *Applied Physics Letters,* vol. 89, pp. -, Sep 4 2006.

[6] Z. Xu, S. H. Li, L. Ma, G. Li, and Y. Yang, "Vertical organic light emitting transistor," *Applied Physics Letters,* vol. 91, pp. -, Aug 27 2007.

[7] A. G. Rinzler, J. Liu, H. Dai, P. Nikolaev, C. B. Huffman, F. J. Rodriguez-Macias, P. J. Boul, A. H. Lu, D. Heymann, D. T. Colbert, R. S. Lee, J. E. Fischer, A. M. Rao, P. C. Eklund, and R. E. Smalley, "Large-scale purification of single-wall carbon nanotubes: process, product, and characterization," *Applied Physics a-Materials Science & Processing,* vol. 67, pp. 29-37, Jul 1998.

[8] Z. C. Wu, Z. H. Chen, X. Du, J. M. Logan, J. Sippel, M. Nikolou, K. Kamaras, J. R. Reynolds, D. B. Tanner, A. F. Hebard, and A. G. Rinzler, "Transparent, conductive carbon nanotube films," *Science,* vol. 305, pp. 1273-1276, Aug 27 2004.

Mater. Res. Soc. Symp. Proc. Vol. 1154 © 2009 Materials Research Society 1154-B05-15

Study of Solid/Liquid Interfaces in Organic Field-Effect Transistors With Ionic Liquids

Ono, Shimpei[1], Miwa, Kazumoto[1]; Seki, Shiro[1]; Takeya, Jun[2]

[1]Central Research Institute of Electric Power Industry, Materials Science Research Laboratory, Komae, Tokyo, Japan
[2]Osaka University, Graduate School of Science, Toyonaka, Osaka, Japan

ABSTRACT

We report high-mobility rubrene single-crystal field-effect transistors with ionic-liquid electrolytes used for gate dielectric layers. As the result of fast ionic diffusion to form electric double layers, their capacitances remain more than 1.0 $\mu F/cm^2$ even at 0.1 MHz. With high carrier mobility of 9.5 cm^2/Vs in the rubrene crystal, pronounced current amplification is achieved at the gate voltage of only 0.2 V, which is two orders of magnitude smaller than that necessary for organic thin-film transistors with dielectric gate insulators. The results demonstrate that the ionic-liquid/organic semiconductor interfaces are suited to realize low-power and fast-switching field-effect transistors without sacrificing carrier mobility in forming the solid/liquid interfaces.

INTRODUCTION

Organic field-effect transistors (OFETs) have attracted much attention because of their potential applications in large-area, flexible, and low-cost electronics [1]. For the development of higher-performance devices, numbers of material combinations are being intensively tested for the layered structure of organic FETs because the interfacial phenomena are crucial in determining their device performances [2]. As one of the unique examples, there has been considerable interest in using electric double layers (EDLs) of electrolytes for efficient application of gate electric field. Since typical thickness of the EDLs is only ~ 1 nm, much higher-density carriers are accumulated at the surface of semiconductor channels than with commonly used 100-500 nm thick SiO_2 or polymer gate dielectrics. Therefore, such devices realize excellent current amplification even at small gate voltage V_G less than 1 V [3]. Recently it has been reported by several groups that the EDL gating technique is indeed useful for high performance OFETs [4-11]. Furthermore, it is also demonstrated for oxide compounds that the high-density carriers introduced by the EDL gating provoke electronic phase transitions from insulating to metallic or even to superconducting phases [12,13], further testifying importance of the electrolyte-gating technique. However, there has not been any detailed study on microscopic mechanisms of the charge transport taking place in the very vicinity of the interfaces to the electrolytes. Since both the transistor performances and properties of the electric phase transitions should be critically influenced by polarization effects of the electrolytes adjacent to the semiconductors, experiments of varying electrolyte materials are desired to develop a detailed description of the electronic states of the interfacial carriers, which is absolutely necessary for both of the above subjects.

In this work we introduce ionic liquids (ILs), also known as room-temperature molten salts, for the electrolyte in the rubrene single-crystals EDL-FETs. Without any solvent, ILs show distinctive properties of high thermal stability and no volatility, therefore attracting interests in the material properties themselves [14]. The advantage of the ILs electrolyte in the OFETs is its high-speed formation of the EDLs; the rapid ionic diffusion of the ions typically in ~ 1 μs is translated into 1 MHz-switching OFETs. We also show that it is possible to keep the surface of the rubrene crystal relatively in a good condition by choosing a proper compound for the ILs, so that mobility of 9.5 cm^2/Vs is realized. As the result, sufficiently large ON current is achieved even with the gate voltage less than 0.5 V, with the benefit of the large capacitance of the EDLs.

EXPERIMENT

The ionic liquids which we use in this experiment are 1-ethyl-3methylimidazolium bis(fluorosulfonyl) imide ([emim][FSI]) and 1-ethyl-3methylimidazolium bis(trifluoromethanesulfonyl) imide ([emim][TFSI]); the structure of these materials are shown in Fig. 1(a). These ILs in the imidazolium family have been known to have rather high ion-conductivity at room temperature (~ 2 - 3×10^{-2} S/cm) and widely used in the studies for potential application in a variety of solid state electrochemical devices, such as fuel cells [15], and lithium batteries [16,17]. All the ILs is commercial ones. Purity of the ILs was checked by the 1H-NMR measurements and Differential Scanning Calorimetry (DSC) measurements. From these measurements, we did not observe any indication of impurities.

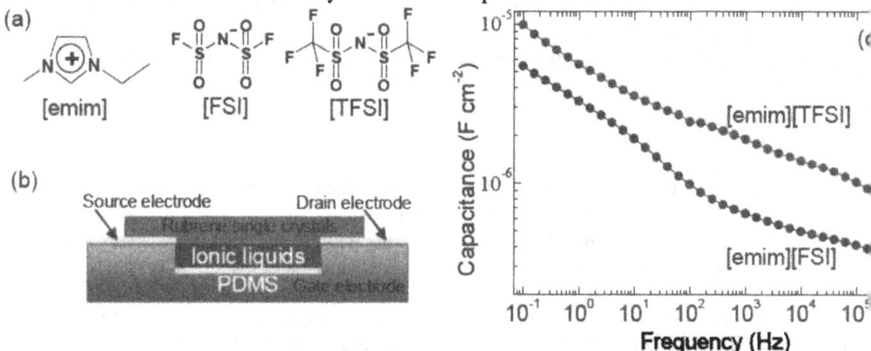

Figure 1. (a) Chemical structures of the ionic liquids; [emim] (left) as cation and [FSI] (center) and [TFSI] (right) as anion. Ionic liquids used in these devices are the combination of cation and anion. (b) Structure of organic crystal / ionic liquid transistors. (c) Capacitance of the ionic liquids as a function of frequency measured by the ac impedance technique.

We have formed a well structure of polydimethylsiloxan (PDMS) elastomer as the substrate on which rubrene single crystal is electrostatically attached as shown in Fig.1(b) [11]. The electrodes of gate, source and drain are all prepared with gold and the gap layer is typically as thick as 25 μm. The ILs are poured underneath rubrene single crystals by the capillary force, so that the EDLs in the ILs can induce high-density carriers at the surface of the crystal with the application of minimum V_G. We employ four-terminal methods to measure the transfer characteristics of the device [18]. All the measurements are done in air at room temperature with an Agilent Technology B1500A semiconductor parameter analyzer.

EDL capacitances C_{EDL} of the ILs are measured using a test device consisting of the same PDMS substrates, where the top surface is structured with gold evaporated on the PDMS substrate. Solartron 1260 and 1296 impedance analyzers are used to obtain the frequency profiles over the range from 0.1 Hz to 1 MHz with the application of AC voltage amplitude of 5 mV.

RESULTS AND DISCUSSION

Fig. 1(c) shows the frequency dependence of capacitance of ILs. The values of the capacitance increase with decreasing frequency to reach 11 $\mu F/cm^2$ and 5.4 $\mu F/cm^2$ at 0.1 Hz for [emim][TFSI] and [emim][FSI], respectively. Though the reason of the frequency dependence in the range of 0.1-10 Hz has not been elucidated yet, it is suspected that rotational rearrangement of the ionic molecules is responsible for the observation of the slow relaxation. As a consequence of such high values of the EDL of [emim][TFSI], application of the gate voltage of 0.5 V leads to the carrier density as high as ~1.0□$10^{13}/cm^2$, which is comparable to the maximum carrier density that the usual SiO_2 devices can reach. It is to be emphasized that the EDL capacitance of the ILs remain large even at 1 MHz: the value is only one order of magnitude lower than that at 0.1 Hz for the both compounds, demonstrating the fast ionic diffusion in response to the voltage application. Therefore, the EDL OFETs incorporating the ILs allow switching operation at such a high frequency, which none of the previously reported devices with polymer electrolytes do.

Before we introduce the ILs into the device, the rubrene single crystals are characterized by measuring transistor performance with the application of gate voltage to the 25 μm air-gap layer. Figure 2(a) shows transfer characteristics of the air-gap transistor. The hole mobility μ is estimated from the transfer characteristics by employing the standard formula in the Ohmic regime $\mu = 1/C\ d\sigma/dV_G$, where channel conductivity σ is given as $\sigma = L/w\ I_D/\Delta V$ with C, I_D, ΔV, L, w being capacitance, drain current, potential drop between the electrode in the middle of the source and drain electrodes, channel length 40 μm, and channel width 150 ~ 200 μm, respectively. As the result, the values of μ are in the range of 10 - 37 cm^2/Vs with the average values around 20 cm^2/Vs, which is comparable to the best value ever reported for the rubrene single-crystal transistors [19,20].

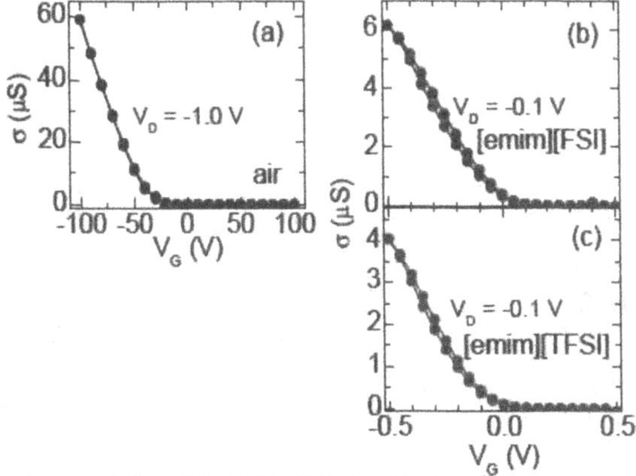

Figure 2. Transfer characteristics of the EDL OFETs with air-gap (a), [emim][FSI] (b), and [emim][TFSI] (c).

Figure 2(a) and (b) show transfer characteristics of the device with [emim][FSI] and [emim][TFSI], respectively. Simultaneously measured gate leakage current through the electrolytes is negligibly small as compared to the drain current I_D, that is less than 0.1 nA as long as $|V_G|$ is less than 0.2 V. A sweeping rate of 0.1 V/s results in negligible hysteresis for both ILs and is translated to the frequency of 10 Hz to change 10 mV, which is the amplitude of the capacitance measurement shown in Fig.1(c). Assuming the values of C_{EDL} at 10 Hz, the average mobility values are 3.7 cm^2/Vs for [emim][TFSI], and 9.5 cm^2/Vs for [emim][FSI]. We emphasize that more than 20 devices are measured for both IL to confirm the reproducibility. These values are relatively high in the standard of ever reported for EDL-gated FETs, of course partially due to the use of the rubrene single crystals with high intrinsic mobility. Also the mobility values for the EDL transistors are comparable to those for the SiO_2 devices, the ILs provide efficient coupling to the organic single crystals, without sacrificing carrier mobility at the solid/liquid interface between organic semiconductors and the ILs.

Figure 3 shows output characteristics of the rubrene crystal FETs with the two IL electrolytes. It is noticeable that rather large current in the order of 100 nA is generated with the application of less than 0.2 V for both V_G and V_D, demonstrating the very low-power operation of the organic transistors due to the high density carrier accumulation at the IL / rubrene liquid-to-solid interfaces. Again relatively small hysteresis are observed for all ILs.

As for the difference in μ of the two ILs, it can be argued that the surface charge is dressed with differently polarizable gate dielectrics; Although the effect of the liquid electrolytes should be different from that of solid gate insulators, the present results qualitatively follow the tendency that μ increases with gate insulators of smaller ε, as reported for carriers on common dielectric insulators [21,22]. On the other hand, Shimotani et al. explain that the scattering events of the carriers by the ion are relevant in the electrolyte [6]. Including our present results on the IL / rubrene semiconductor interfaces, it is to be further investigated to elucidate microscopic mechanisms of the charge transport in the vicinity of the solid/liquid interfaces.

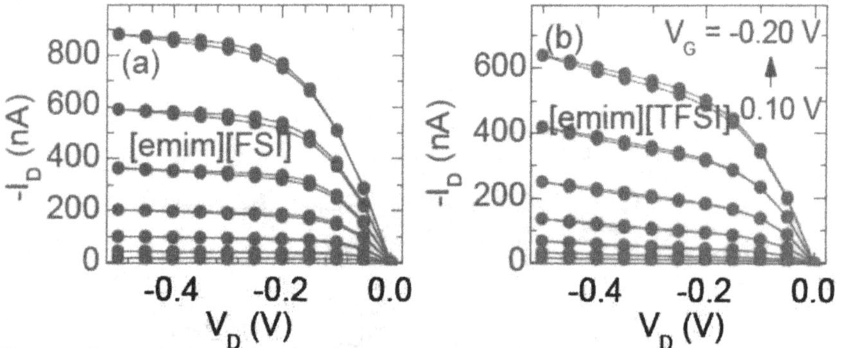

Figure 3. Output characteristics of the devices with [emim][FSI] (a) and [emim][TFSI] (b) with different gate voltages (V_G = 0.10, 0.05, 0.00, -0.05, -0.10, -0.15, -0.20 V, respectively).

CONCLUSIONS

In conclusion, it is demonstrated that the use of ionic-liquid electrolytes in organic single-crystal FETs enable high-density carrier doping with minimum gate voltages without sacrificing the carrier mobility. Therefore, the ionic-liquid gating can be a promising technology to realize high-mobility, fast-switching and low-power organic transistors. Owing to varieties of ILs that have been synthesized so far, higher performances are likely to emerge by elaborate search for compounds incorporated in organic FETs. Due to the capability of the high-density carrier doping, the present technique is also useful in more basic material sciences such as carrier-density driven phase transition in strongly correlated electron systems.

ACKNOWLEDGMENTS

The authors would like to thank M. Tominari for his technical assistance and I. Tsukada for helpful discussions. This study was supported in part by Grants-in-Aid for Scientific Research (Nos. 17069003, 18028029,19360009 and 20740213) from the Ministry of Education, Culture, Sports, Science, and Technology, Japan.

REFERENCES

1. G. Malliaras and R. Friend, *Phys. Today.* **58**, 53 (2005).
2. J. Veres, S. D. Ogier, S. W. Leeming, D. C. Cupertino, and S. M. Khaffaf *Adv. Funct. Mater.* **13**, 193 (2003).
3. M. J. Panzer and C. D. Frisbie, *J. Am. Chem. Soc.* **127**, 6960 (2005).
4. M. J. Panzer and C. D. Frisbie, *Appl. Phys. Lett.* **88**, 203504 (2006).
5. J. Takeya, K. Yamada, K. Hara, K. Shigeto, K. Tsukagoshi, S. Ikehata, and Y. Aoyagi, *Appl. Phys. Lett.* **88**, 112102 (2006).
6. H. Shimotani, H. Asanuma, J. Takeya, and Y. Iwasa, *Appl. Phys. Lett.* **89**, 203501 (2006).
7. E. Said, X. Crispin, L. Herlogsson, S. Elhag, N. D. Robinson, and M. Berggren, *Appl. Phys. Lett.* **89**, 143507 (2006).
8. J. Lee, M. J. Panzer, Y. He, T. P. Lodge, and C. D. Frisbie, *J. Am. Chem. Soc.* **129**, 4532 (2007).
9. M. J. Panzer and C. D. Frisbie, *J. Am. Chem. Soc.* **129**, 6599 (2007).
10. S. Ono, S. Seki, R. Hirahara, Y. Tominari, and J. Takeya, *Appl. Phys. Lett.* **92**, 103313 (2008).
11. T. Uemura, R. Hirahara, Y. Tominari, S. Ono, S. Seki, and J. Takeya, *Appl. Phys. Lett.* **93**, 263305 (2008).
12. H. Shimotani, H. Asanuma, A. Tsukazaki, A. Ohtomo, M. Kawasaki, and Y. Iwasa, *Appl. Phys. Lett.* **91**, 082106 (2007).
13. K. Ueno, S. Nakamura, H. Shimotani, A. Ohtomo, N. Kimura, T. Nojima, H. Aoki, Y. Iwasa, and M. Kawasaki, *Nature Mat.* **7**, 855 (2008).
14. R. Misra, M. McCathy, and A. F. Hebard, *Appl. Phys. Lett.* **90**, 052905 (2007).
15. M. A. B. H. Susan, T. Kaneko, A. Noda, and M. Watanabe, *J. Am. Chem. Soc.* **127**, 4976 (2005).
16. S. Seki, Y. Ohno, Y. Kobayashi, H. Miyashiro, A. Usami, Y. Mita, H. Tokuda, M. Watanabe, K. Hayamizu, S. Tsuzuki, M. Hattori, and N. Terada, *J. Electrochem, Soc.* **154**, A173 (2007).
17. S. Seki, Y. Ohno, Y. Kobayashi, H. Miyashiro, A. Usami, Y. Mita, H. Tokuda, M. Watanabe, K. Hayamizu, S. Tsuzuki, M. Hattori, and N. Terada, *J. Electrochem, Soc.* **154**, A173 (2007).
18. J. Takeya, J. Kato, K. Hara, M. Yamagishi, R. Hirahara, K. Yamada, Y. Nakazawa, S. Ikehata, K. Tsukagoshi, Y. Aoyagi, T. Takenobu, and Y. Iwasa, *Phys. Rev. Lett.* **98**, 196804

(2007).
19. E. Menard, V. Podzorov, S. H. Hur, A. Gaur, M. E. Gershenson, and J. A. Rogers, *Adv. Mater.* **16**, 2097 (2004).
20. J. Takeya, M. Yamagishi, Y. Tominari, R. HIrahara, Y. Nakazawa, T. Nishikawa, T. Kawase, T. Shimoda, and S. Ogawa, *Appl. Phys. Lett.* **90**, 102120 (2007).
21. A. F. Stassen, R. W. I. de Boer, N. N. Iosad, and A. F. Morpurgo, *Appl. Phys. Lett.* **85**, 3899 (2004).
22. I. N. Hulea, S. Fratini, H. Xie, C. L. Mulder, N. N. Iossad, G. Rastelli, S. Ciuchi, and A. F. Morpurgo, *Nature Mat.* **5**, 982 (2006).

Mater. Res. Soc. Symp. Proc. Vol. 1154 © 2009 Materials Research Society 1154-B05-23

Single and Mixed Self-Assembled Monolayers of Phenyl Species on SiO_2 With Various Ring to Ring Interactions

Virginie Gadenne [1,2,3], Simon Desbief [1,2,3] and Lionel Patrone [1,2,3]
[1] Aix-Marseille Université, IM2NP
[2] CNRS, IM2NP (UMR 6242)
[3] Institut Supérieur de l'Electronique et du Numérique, IM2NP
Maison des Technologies, Place Georges Pompidou, F-83000 Toulon, France

ABSTRACT

We studied the formation of phenylalkyltrichlorosilane self-assembled monolayers on native oxide covered silicon. After a first chemisorption step in the monolayer growth, the presence of the short alkyl chain (3-4 carbon atoms) is responsible for a second growth step which corresponds to the arrangement between molecules. We found that this packing step is accelerated by replacing phenyl by pentafluoro-phenyl rings, possibly due to quadrupolar interactions between fluorinated cycles. Furthermore we demonstrate that mixing phenyl and pentafluoro-phenyl molecules leads to an even faster packing step which is accounted for by hydrogen bonding CH...FC in a face to face phenyl/pentafluoro-phenyl arrangement. We believe these results allow improving charge delocalization over conjugated molecular domains. In a second part, we studied the phase separation between phenyl-alkyltrichlorosilane and octadecyltrichlorosilane (OTS) molecules. Improving the phase separation was studied using ring to ring interactions afore-analyzed. We show phase separation is improved and OTS islands are smaller with phenyl species that involve stronger ring to ring interactions. The best case is obtained with mixing phenyl and pentafluoro-phenyl rings using hydrogen bonds for packing together the aromatic species. These results demonstrate improved control of SAM composition and morphology essential to further use the obtained islands for building molecular devices.

INTRODUCTION

Molecular self-assembly [1,2], which consists in the adsorption of molecules on a substrate within a spontaneously ordered monolayer, is one of the most promising issues for giving surface specific properties. An important field of application of self-assembled monolayers (SAMs) [2] is molecular electronics within which self-assembly is a very powerful way to obtain the organization at large surface scale of molecules showing particular electronic properties (insulator, molecular wire, memory, diode, ...). Beside the numerous studies devoted to molecular SAMs on metal surfaces such as thiols on gold [2], using silicon as a SAM substrate appears to be a promising challenge in molecular electronics. Indeed, this hybrid approach benefits from both the well-developed silicon technology and the specific properties of molecules. Therefore, preparation of self-assembled monolayers (SAM) of aromatic conjugated molecules on silicon is a key point in molecular electronics [3]. Moreover, regarding potential applications, it is important to be able to prepare nano-islands of such active molecules on silicon. Nevertheless few works addressed this subject [4]. For these two points, strong interactions between molecules are helpful. Within conjugated SAMs, in order to obtain a charge carrier delocalization in the plane, aromatic cycles must be packed together [5]. Concerning the formation of small aromatic domains, it is expected it can be achieved using phase separation between different species. One way to improve the phase separation is to increase the interactions at least between species forming one of the two phases. This is the scope of this work using two aromatic molecules bearing a phenyl ring on a small alkyl chain. One of the two studied molecules owns fluorine atoms in peripheral position around the phenyl ring. In this case, mixing the two molecules allows introducing hydrogen bonding between the phenyl rings [6], which is expected to increase the packing of the aromatic SAM and to favor the phase separation with a third kind of molecule such as an alkyl chain. This study was carried out on silicon covered with its natural oxide using phenylalkyl molecules ended by a trichlorosilane reactive head for this purpose [7].

EXPERIMENT

Molecules

Aromatic molecules chosen to perform this study are phenylbutyltrichlorosilane (referred to as PBTCl) and pentafluoro-phenyltrichlorosilane (referred to as FPPTCl) which are presented in Figure 1. The one carbon-length difference between the two alkyl chains of these molecules allows compensating the longer C-F bonds and thus promoting CH...FC hydrogen bonding between phenyl rings in the SAM of the mixed molecules [8]. In order to test the strength of the interaction between phenyl rings, we used octadecyltrichlorosilane (referred to as OTS) as a third molecule which is able to growth within islands at low enough temperature [9].

Figure 1. Aromatic trichlorosilane molecules used in this study: (a) phenylbutyltrichlorosilane (PBTCl), ~12 Å in length, (b) pentafluoro-phenyltrichlorosilane (FPPTCl), ~11 Å in length.

Deposition protocol

The substrates are cut from Si (100) wafers covered with native oxide. First, the substrate is degreased in a sonicated chloroform bath, and then dried under a nitrogen flow. The substrate is then dipped into a piranha mixture (H_2SO_4 / H_2O_2, 7/3, *caution!*) for 30 minutes at 150 °C in order to remove any organic impurities from the surface and to increase the amount of hydroxyl moieties (OH) necessary for the grafting of silane heads. It is then rinsed abundantly with pure de-ionized water (18 MΩ.cm) and quickly immersed into a de-ionized water beaker. Still in this beaker, the substrate is then transferred into a glove-box filled with nitrogen at 40% relative humidity in which silanization is performed. At this step, the silicon is covered with a clean native thin oxide with an OH-rich surface. The substrate is dried under a nitrogen flow and dipped into the silanization solution, consisting of a mixture of a linear alkane (hexadecane, 99+%), carbon tetrachloride (CCl_4) and the trichlorosilanes (at 10^{-2} M). This solution was beforehand thermalized on a thermostated plate during ~20 minutes at 11°C to which it is kept during all the silanization time of 1h30min. Then, the sample is rinsed in a sonicated chloroform bath and dried under a nitrogen flux.

Ellipsometry, contact angle measurements, and Atomic Force Microscopy (AFM)

We first measure the monolayer thickness by ellipsometry to check the SAM is uniformly complete. Moreover, this measurement is the first appreciation of the composition of the mixed OTS/aromatic SAMs. Indeed, due to the large length difference between OTS (length around 26 Å) and aromatic molecules (11-12 Å in length), the measured thickness allows to estimate the proportion of OTS in the SAM. The measurements were carried out using a *Sentech SE400*, with a 632.8 nm He-Ne laser at an incidence angle of 70°. Taking the value of oxide thickness as a zero, we could measure the SAM thickness assuming an optical index of n=1.45 for OTS [10] and n=1.50 for aromatic molecules [7] (considering the material isotropic and homogeneous). All the thickness values are an average calculated from at least 5 measurements over 3 samples.

Water static contact angles are measured to appreciate the homogeneity of the monolayer. In addition, it gives us another estimation of the composition of OTS/aromatic mixed SAMs using Cassie's law [11]: if "θ" is the angle measured on the mixed SAM, "a" and "b" the proportion of the two types of molecules in the SAM having respective contact angles "θ_a" and "θ_b", then $\cos\theta = a.\cos\theta_a + b.\cos\theta_b$. OTS single SAMs show a typical water contact angle of 110° whereas aromatic PBTCl and FPPTCl SAMs exhibit angles of ~86° and ~90° respectively. Our measurements were made using a *DSA 10 MK2* from *Krüss GmbH.* All the given angles are an average calculated from at least 5 measurements over 3 samples.

AFM measurements are performed using a Multimode system equipped with a *Nanoscope IIIa* controller from *Veeco Instruments Inc.* All images are recorded under ambient conditions, in Tapping™ AFM mode, with Si tips with a resonance frequency of 150-300 kHz. AFM images allow us to measure the covering rate of the structures present on the surface. The measure of the island coverage of OTS has been made using the flooding function of the *WSxM* software [12]. By choosing the right threshold level, the program gives an estimation of the coverage of structures above this level.

DISCUSSION

Growth kinetics of PBTCl, FPPTCl and PBTCl:FPPTCl (1:1) SAMs measured using water contact angle are presented in Figure 2.

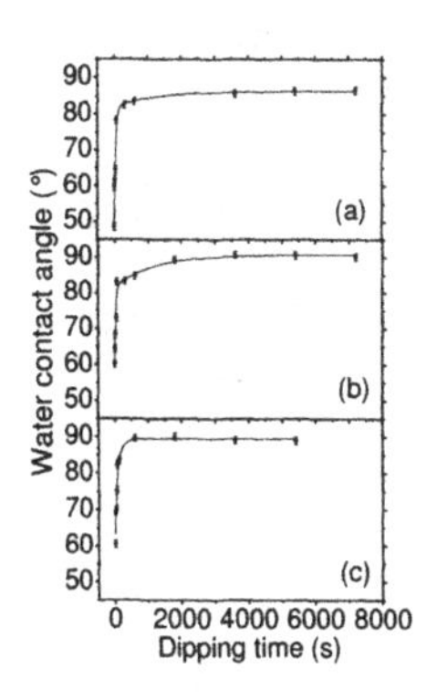

Figure 2. Growth kinetics of SAMs of (a) PBTCl, (b) FPPTCl, (c) PBTCl:FPPTCl (1:1) measured from the water contact angle. Solid line is the best fit carried out using equation 1.

They are best-fitted using a Langmuir-type growth function with two time constants t_1 and t_2:

$$A = A_{max}[1 - C_1.\exp(-t/t_1) - C_2.\exp(-t/t_2)] \qquad (1)$$

Time constants are reported in Table I. It can be noticed that t_1 is similar for the three cases: it may therefore correspond to the step of molecule chemisorption on the surface since it is expected to depend only on the grafting head which is the same for both molecules. Moreover for OTS that does not bear any phenyl moiety we measured a single step growth with a time constant around 35 seconds [13], very close to the values of t_1 reported in Table I. Concerning the second step, the time constant t_2 is larger, being consistent with a densification process which occurs through rearrangement between the top aromatic moieties. This step is shown to be strongly accelerated for the mixing of PBTCl and FPPTCl as t_2 value falls down. Such a behavior can be explained by the nature of the interactions involved between aromatic rings: π-π for PBTCl, quadrupolar for FPPTCl [8] and hydrogen type for their mixing [14,15]. These results show that the stronger the interactions between phenyl rings the more important the involved long range forces thus explaining a quicker molecular rearrangement within the SAM.

Table I. Time constants from the fit by equation (1) of the growth kinetics presented in Figure 2.

	t_1 (s)	t_2 (s)
PBTCl	42	1188
FPPTCl	30	1164
PBTCl+FPPTCl	29	156

In order to test the influence of the various interactions involved within these aromatic SAMs and the possibility they offer to obtain small nano-domains of a third molecule, we studied the SAMs prepared from the mixing of the previous molecules with OTS. At the silanization temperature used (11°C), it is expected OTS should undergo an island growth [9] whereas the small aromatic molecules should growth in a disordered phase before their rearrangement occurring in a second step [16]. Mixed OTS /aromatic SAMs were prepared with various ratios ranging from 0.2 to 10 using PBTCl, FPPTCl or PBTCl:FPPTCl (1:1) as aromatic molecules. AFM images of the SAMs prepared with a ratio of 2 are presented in Figure 3. One can see the presence of molecular islands protruding from the surrounding phase. The height of these islands measured from the cross sections is consistent with the length difference between OTS and aromatic molecules. As a consequence, the islands should be made mainly of OTS molecules inside a phase mainly composed of aromatic molecules.

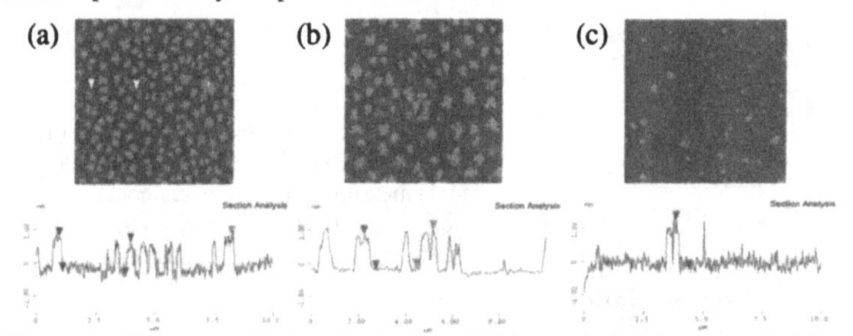

Figure 3. AFM images (5x5 µm²) of mixed OTS/aromatic SAMs prepared with a ratio of 2: (a) OTS/PBTCl, (b) OTS/FPPTCl, (c) OTS/PBTCl:FPPTCl (1:1). Cross-sections have been made

on a selection of islands showing their height is compatible with the one of OTS islands protruding from a phase mainly composed of aromatic molecules.

It appears clearly from AFM images that in the case of mixed PBTCl:FPPTCl aromatic molecules the number and the size of OTS islands decrease while the aromatic phase covers more surface. Such an evolution can be related to the increasing strength of the intermolecular forces involved within the aromatic phase of these SAMs. The stronger they are, the more it is difficult for OTS molecules to diffuse on the surface and to gather within islands. Mixing PBTCl with FPPTCl should favor a face to face arrangement of their respective rings due to CH...FC hydrogen bonding which occurs between the five CH and CF moieties all around the rings. This case should involve stronger interactions than between phenyl rings within PBTCl or FPPTCl single aromatic phase.
In order to estimate the amount of OTS and aromatic molecules in the SAM prepared with all the ratios, we performed ellipsometry and water contact angle measurements and we extracted the island coverage from AFM images. From these measurements we assessed an OTS amount in the SAM that we plotted as a function of OTS amount introduced in the solution. Results are presented in Figure 4.

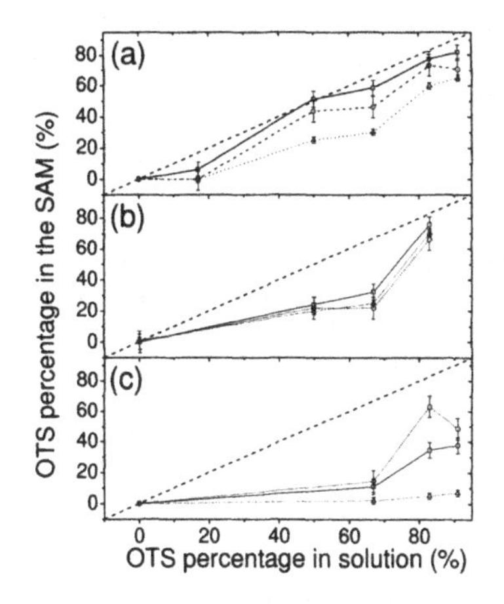

Figure 4. OTS proportion estimated in the mixed OTS/aromatic SAM from ellipsometry (○), water contact angle (□) and island coverage (△), as function of OTS proportion injected in the solution, for (a) PBTCl, (b) FPPTCl, (c) PBTCl:FPPTCl (1:1) aromatic molecules. Dashed line represents the case where OTS proportion in the SAM is equal to the one in solution.

In these plots, the better the agreement between the OTS island coverage and the OTS amount estimated from ellipsometry and water contact angle, the better the phase separation between OTS and the aromatic molecules. The amount of OTS in the SAM is close to the one injected in the solution for PBTCl (Figure 4a) and the best phase separation is obtained for FPPTCl (Figure 4b). As a tentative explanation, possible repulsion between CF moieties of the fluorinated phenyl rings and CH_2 groups of OTS molecules may be invoked as observed between alkyl chains and perfluoro-alkyl chains leading to phase separation [16]. On the contrary, stronger ring to ring interactions with PBTCl/FPPTCl mixing appear not to be the best conditions for phase separation. In this case, the more strongly packed aromatic phase may trap OTS molecules preventing them from diffusing to gather together.

CONCLUSIONS

We have shown that the growth on silicon dioxide of SAMs of short phenyl-alkyltrichlorosilane species exhibits two steps: chemisorption depending on the grafting head, and a longer step of densification depending on the interactions involved between phenyl rings. This second step is about height times quicker by introducing hydrogen bonding between phenyl rings while mixing phenylbutyltrichlorosilane and pentafluoro-phenylpropyltrichlorosilane molecules. Moreover, the interactions between aromatic rings in the monolayer modify the composition of the final SAM prepared with OTS. These interactions between phenyl rings impact on the size and quantity of alkyl nano-domains and moderate interactions can improve phase separation with the alkyl chains. Further work is addressed to improve this control. In particular, mixing molecules with reactive moieties having different grafting kinetics may offer another possibility to control the SAM structure and composition.

ACKNOWLEDGMENTS

This work was performed and presented within the framework of the world class competitive cluster "Secured Communicating Solutions (SCS)" that is acknowledged for financial support. Equipment was mainly funded by the "*Objectif 2*" EEC program (FEDER), the "*Conseil Général du Var*" Council, the PACA Regional Council and Toulon Provence Méditerranée which are acknowledged.

REFERENCES

1. G.M. Whitesides, B. Grzybowski, Science **295**, 2418 (2002).
2. A. Ulman, *An introduction to ultrathin organic films* (Academic Press: Boston, 1991)
3. M. Halik, H. Klauk, U. Zschieschang, G. Schmid, C. Dehm, M. Schütz, S. Maisch, F. Effenberger, M. Brunnbauer, F. Stellacci, Nature **431**, 963 (2004)
4. F. Fan, C. Maldarelli, A. Couzis, Langmuir **19**, 3254 (2003)
5. J. Collet, S. Lenfant, D. Vuillaume, O. Bouloussa, F. Ro,delez, J.M. Gay, K. Kham, C. Chevrot, Appl. Phys. Lett. **76**, 1339 (2000)
6. J.D. Dunitz, ChemBioChem **5**, 614 (2004)
7. J. Moineau, M. Granier, G.F. Lanneau, Langmuir **20**, 3202 (2004).
8. E.A. Meyer, R.K. Castellano, F. Diederich, Angew. Chem. Int. Ed. **42**, 1210 (2003)
9. C. Carraro, O.W. Yauw, MM. Sung, R. Maboudian, J. Phys. Chem. B **102,** 4441 (1998)
10. C.D. Bain, G.M. Whitesides, *J. Am. Chem. Soc.* **111**, 7164 (1989)
11. A.B.D. Cassie, S. Baxter, Trans. Faraday Soc. **40**, 546 (1944)
12. Software *WSxM* from *Nanotec Electronica*, freeware www.nanotec.es
13. S. Desbief, L. Patrone, D. Goguenheim, D. Guerin, D. Vuillaume, (submitted)
14. V.R. Thalladi et al., J.Am.Chem.Soc. **120**, 8702 (1998)
15. S. Zhu, S. Zhu, G. Jin, Z. Li, Tetrahedr. Lett. **46**, 2713 (2005)
16. Brzoska, J.B.; Ben Azouz, I.; Rondelez, F. Langmuir **10**, 4367 (1994)
17. W. Mizutani, T. Ishida, S.I. Yamamoto, H. Tokumoto, H. Hokari, H. Azehara, M. Fujihira Appl. Phys. A **66**, S1257-S1260 (1998)

Mater. Res. Soc. Symp. Proc. Vol. 1154 © 2009 Materials Research Society 1154-B05-49

FET Characteristics of Chemically-Modified CNT

R. Kumashiro[1], Y. Wang[1,2], N. Komatsu[1] and K. Tanigaki[1,2]
[1]Department of Physics, Graduate School of Science, Tohoku University,
6-3 Aoba Aramaki, Aoba-ku, Sendai, Miyagi 980-8578, Japan
[2]Advanced Institute for Materials Research - World Premier International Research Center,
Tohoku University,
2-1-1 Katahira, Aoba-ku, Sendai, Miyagi 980-8577, Japan

ABSTRACT

Electric transport properties of chemically modificated carbon nanotubes (CNT) using Si-containing organic molecules and polymers were investigated by means of the field effect transistors (FET) technique. From the results of FET measurements for each chemically surface modified CNT, it was shown that p-type semiconducting CNT can be converted to n-type ones by physical adsorption of Si-containing organic molecules and polymers having Ph-groups. It is suggested that the electron carrier are doped into CNT from the adsorbed molecules and polymers, and it was also confirmed by the results of adsorption spectra. That is, it can be said that the electronic properties of CNT can be controlled by chemically modifications of outer surface.

INTRODUCTION

CNT have a bright prospect as electronic materials for nano-scale devices in the future, and a large number of studies have been made in recent years [1-3]. In particular, it is well known that the FET fabricated from CNT having semiconducting properties show high ability in terms of the mobility [4]. However, carriers in pristine CNT are mostly hole, therefore, CNT-FET usually show the p-type properties [5]. For applying CNT to electronic devices, it is necessary to control the carriers of both electrons and holes, that is, the electron carrier doping should be established. As electron carrier doping techniques for CNT, three major techniques are generally possible, substitution by hetero carbon atoms, endohedral doping, and exohedral modification. Recently, it has reported that the structural substitution of CNT by hetero carbon atoms, such as group III or V element, changes the electronic properties of CNT [6]. It has been also known that the doping into inner spaces of CNT with alkali metals or organic molecules can change the properties of CNT-FET [7, 8]. Exohedral modifications for CNT by organic reaction have been carried out for the purpose of purification, solubilization, functionalization, and so on [9, 10]. However, a small number of studies have been made relating to the investigation of physical properties. A similar carrier doping could exohedrally be possible when the SWNT surface is chemically modified. With such chemical modifications, the charge transfer from the substituent groups to SWNT will be expected and this could modify the electronic states of SWNT.

Recently, we have reported the FET properties of individual SWNT exohedrally modified by Si-containing organic moieties, and demonstrated that p-type nanotubes can be converted to n-type ones [11, 12]. However, because of ununiformity of the surface-chemical modifications of SWNTs, the true effects of the exohedral modifications on FET properties were extremely difficult to be evaluated. In this study, we have made a chemical modification of CNTs using a Si-containing organic molecules and Si-containing organic polymers, and have examined how the FET properties of these chemically modificated CNTs will change.

EXPERIMENT

The CNT samples having single walled structure used in this study were synthesized through the cobalt-molybdenum-catalyst (CoMoCat) processes (South West Nanotechnologies). The diameter of as-prepared CNTs was estimated to be 0.76 to 0.92 nm from the characteristics of fluorescence spectra [13]. Exohedral silylation of CNTs was carried out by a physical adsorption with a silicon containing organic molecule $(SiPh_2{}^tBu)_2$ $(SiPhMe)_n$ polymer and

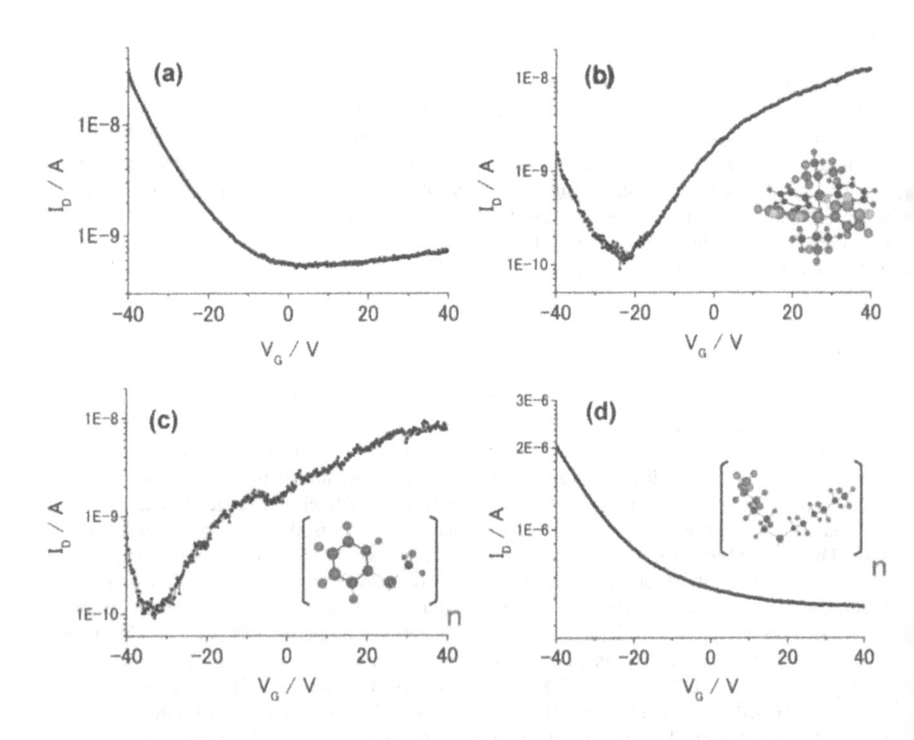

Figure 1. Transfer curves of pristine and Si-containing organic molecule adsorbed CNT-FET: (a) pristine; (b) $(SiPh_2{}^tBu)_2$ molecule adsorbed; (c) $(SiPhMe)_n$ polymer adsorbed; (d) $(n\text{-}Hex_2Si)_n$ polymer adsorbed.

$(n\text{-}Hex_2Si)_n$ polymer in an organic solvent. A highly doped Si wafer was used as a substrate. The gate dielectric is a 300 nm-thick layer of thermally grown SiO_2 film. Au/Ti electrodes in 50 nm thick were used as the source and the drain electrodes. The channel lengthes of the FETs were 10μm. The silylated CNT samples were prepared by supersonic treatment in a THF solvent. The dispersed solution was dropped and spin-coated on the FET substrate. For measurements of FET properties, the CNT-FET were transferred into a high-vacuum chamber and I_{SD}-V_G measurements were carried out by using a micro prober measurement system (Izumi Tech. Co., Ltd) and a semiconductor parameter analyzer (Agilent Technologies, 4155C) at ambient temperature. All measurements were carried out at the condition of 3.0 V of the V_{SD} value. The measurements of absorption spectra were carried out by using the HITACHI U-3410 spectrophotometer in 400-1800 nm wavelength region. As a reference, the experiments were also made in the same manner on chemically non-modified CNT

RESULTS AND DISCUSSION

Fig.1 shows the transfer curves of the spread-sheet CNT-FET before and after silylation. In the case of pristine CNT-FETs (Fig.1a), an ambipolar characteristic, showing a strong p-type semiconducting property, had been observed [11, 12]. This result indicates that the carriers in pristineCNTs are mainly holes. On the contrary, the transfer curves for CNT-FET modificated by the Ph-group containing organic molecules and polymer show strong n-type semiconducting characteristics (Fig.1b,c). In addition, CNT modificated by Si-containing polymer without Ph-group, $(n\text{-}Hex_2Si)_n$ polymer, dose not show strong n-type semiconducting characteristics (Fig.1d). From these results, it can be assumed that the silylated CNTs have electron carriers and p-type CNTs can be converted to n-type ones by exohedral silylation with Ph-group. From these results, it can be also concluded that the n-type CNT-FETs are more produced by the surface silylation, that is, electron carriers are transferred from the Si-containing organic molecule and polymer into CNTs through the π-π interaction between CNT surface and Ph-groups.

Fig.2 shows the results of absorption spectra of pristine and $(SiPh_2{}^tBu)_2$ adsorbed CNT samples. The difference spectrum of pristine and $(SiPh_2{}^tBu)_2$ adsorbed CNT shows the shift of absorption peaks before and after silylation (Fi.2a). This result suggests that the electronic state of CNT was changed by adsorption of $(SiPh_2{}^tBu)_2$ molecules on CNT surface. From the curve analysis of absorption peaks in each spectrum (Fig.2b,c), it was shown that the relative intensities of the peaks in higher-wavenumber region were decreased after adsorption of $(SiPh_2{}^tBu)_2$ molecules on CNT. It can be clearly seen from the comparison of the normalized absorption peak intensities between pristine and $(SiPh_2{}^tBu)_2$ adsorbed samples (Fig.2d). The absorption peaks around 1000 nm were assigned to the E11 transition of the (6,5) CNT having semiconducting nature [13]. Therefore, the decrease in intensities of the absorption peaks by $(SiPh_2{}^tBu)_2$ molecule adsorption suggest that the part of E11 transition were inhibited by the electron filling of the lowest conduction band of semiconductor CNT. That is to say, it is suggested that the electron carrier are doped into CNT from the adsorbed $(SiPh_2{}^tBu)_2$ molecules, and it also support the FET results.

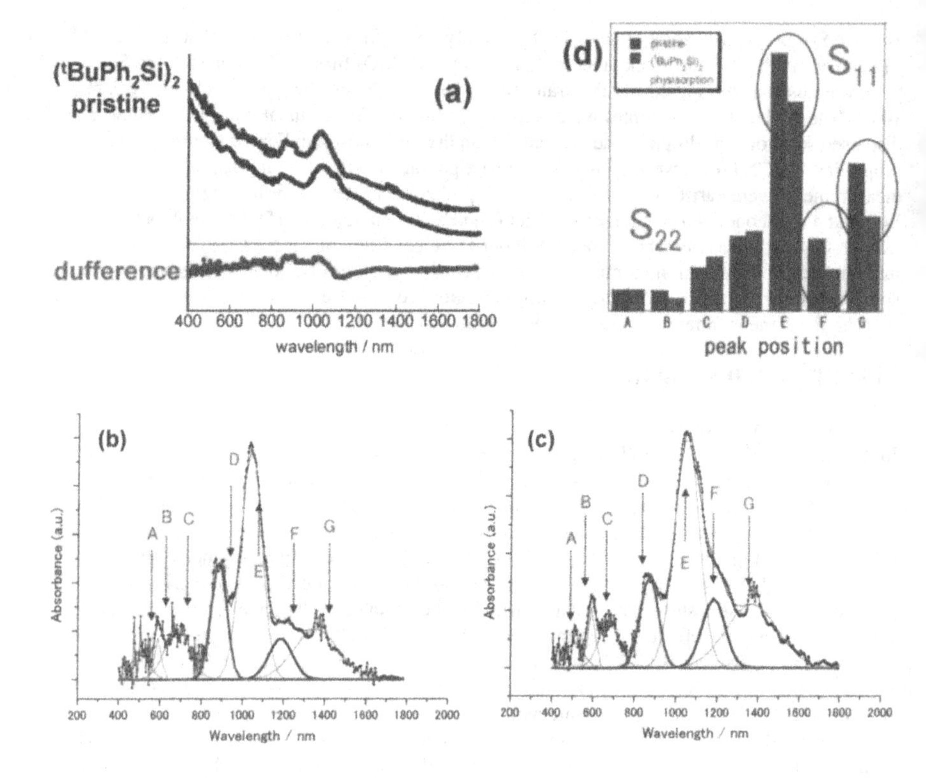

Figure 2. Analysis of absorption spectra of spread-sheeted pristine and $(SiPh_2{}^tBu)_2$ adsorbed CNT samples: (a) absorption spectra of pristine and $(SiPh_2{}^tBu)_2$ adsorbed CNT samples; (b) curve fitting of absorption peaks for pristine sample; (c) curve fitting of absorption peaks for $(SiPh_2{}^tBu)_2$ adsorbed sample; (d) comparison of the normalized absorption peak intensities between pristine (red) and $(SiPh_2{}^tBu)_2$ adsorbed (blue) samples.

CONCLUSIONS

Electric transport properties of chemically modificated CNT using Si-containing organic molecules and polymers were investigated by means of the FET technique. From the results of FET measurements for each chemically surface modified CNT, it was shown that p-type semiconducting CNT can be converted to n-type ones by physical adsorption of Si-containing organic molecules and polymers having Ph-groups. It is suggested that the electron carrier are doped into CNT from the adsorbed molecules and polymers, and it was also confirmed by the results of adsorption spectra. That is, it can be said that the electronic properties of CNT can be controlled by chemically modifications of outer surface.

ACKNOWLEDGMENTS

We thank Prof. Y. Maeda of the Tokyo Gakugei Univ. and Prof. T. Akasaka of the Univ. of Tsukuba for thier assistance in preparing the organic-molecule adsorbed CNT samples. We also thank Prof. N. Kobayashi of the Tohoku Univ. for his assistance in measuring the absorption spectra. This work was supported by Creation of Nano Devices and Systems Based on New Physical Phenomena and Functional Principles of CREST of JST. The present work is also partially supported by the Tohoku University 21 century COE program "Particle-Matter Hierarchy" of MEXT, Japan. This work was performed by a Grant-in-Aid from the Ministry of Education, Science, and Culture of Japan, No.17038801, 17710088, 18651075 and 18204030. This work was carried out under KAKENHI (Grant-in-Aid for Scientific Research) on Priority Areas "New Material Science Using Regulated Nano Spaces-Strategy in Ubiquitous Elements" from the Ministry of Education, Science, and Culture of Japan.

REFERENCES

1. J. Kong, N. R. Franklin, C. Zhou, M. G. Chapline, S. Peng, K. Cho and H. Dai, *Science*, **287**, 622 (2000).
2. A. Bachtold, P. Hadley, T. Nakanishi and C. Dekker, *Science*, **294**, 1317 (2001).
3. V. Derycke, R. Martel, J. Appenzeller and Ph. Avouris, *Nano Lett.*, **1**, 453 (2001).
4. T. Durkop, S. A. Getty, E. Cobas and M. S. Fufrer, *Nano Lett.* **4**, 35 (2004).
5. R. Martel, T. Schmidt, H. R. Shea, T. Hertel and Ph. Avouris, *Appl. Phys. Lett.*, **73**, 2447 (1998).
6. K. Xiao, Y. Liu, P. Hu, G. Yu, Y. Sun and D. Zhu, *J. Am. Chem. Soc.*, **127**, 8614 (2005).
7. C. Zhou, J. Kong, E. Yenilmez and H. Dai, *Science*, **290**, 1552 (2000).
8. T. Takenobu, T. Takano, M. Shiraishi, Y. Murakami, M. Ata, H. Kataura, Y. Achiba and Y. Iwasa, *Nature Materials*, **2**, 683 (2003).
9. J. Chen, M. A. Hamon, H. Hu, Y. Chen, A. P. Rao, P. C. Eklund and R. C. Haddon, Science, 282, 95 (1998).
10. E. T. Mickelson, C. B. Huffman, A. G. Rinzler, R. E. Smalley, R. H. Hauge and J. L. Margrave, *Chem. Phys. Lett.*, **296**, 188 (1998).
11. R. Kumashiro, H. Ohashi, T. Akasaka, Y. Maeda, S. Takaishi, M. Yamashita, S. Maruyama, T. Izumida, R. Hatakeyama and K. Tanigaki, *Mater. Res. Soc. Symp. Proc.*, **901E,** 0901-Rb21-05 (2006).
12. R. Kumashiro, N. Hiroshiba, N. Komatsu, T. Akasaka, Y. Maeda, S. Suzuki, Y. Achiba, R. Hatakeyama and K. Tanigaki, *J. Phys. Chem. Solid*, **69,** 1206 (2008).
13. Y. Maeda, M. Kanda, M. Hashimoto, T. Hasegawa, S. Kimura, Y. Lian, T. Wakahara, T. Akasaka, S. Kazaoui, N. Minami, T. Okazaki, Y. Hayamizu, K. Hata, J. Lu and S. Nagase, *J. Am. Chem. Soc.*, **128**, 12239 (2006).

Mater. Res. Soc. Symp. Proc. Vol. 1154 © 2009 Materials Research Society 1154-B05-84

Enhanced Performance of Organic Light Emitting Diodes Using LiF Buffer Layer

Omkar Vyavahare[1] and Richard Hailstone[2]

[1]Materials Science and Engineering, Rochester Institute of Technology, Rochester, NY 14623
[2]Center for Imaging Science, Rochester Institute of Technology, Rochester, NY 14623

ABSTRACT

Since the invention of organic electroluminescent devices, a great deal of effort has been made to improve their performance. Reducing the barrier and optimizing charge injection is crucial for efficient and bright Organic Light Emitting Diodes (OLEDs). We report the performance of OLEDs with ITO/TPD/Alq3/Al structure by inserting LiF both at electrode-organic interfaces and organic-organic interface. We elucidate the mechanism of the LiF buffer layer inserted at different interfaces. The device with LiF as a cathode injection layer shows improved luminescence and steeper IV characteristics.

INTRODUCTION

Organic Light Emitting Diodes (OLEDs) have been extensively investigated for improving their performance owing to their potential applications in flat-panel displays and solid-state lighting. These electroluminescent devices have the advantages of being self emitting, consuming low power, having a wide viewing angle, and having a faster switching speed [1].

Electron-hole balance and improved charge injection is necessary for the efficiency, brightness and stability of these devices. The majority of carriers in OLEDs are holes due to their higher mobility and smaller injection barrier [2-3]. Hence, enhanced electron injection is desired for charge balance. To achieve it, low work function metals and metal alloys such as Ca, K, Mg, Mg:Ag, and Li have been tried as cathode materials [4-7], but they have low corrosion resistance and high chemical reactivity with the organic medium. Thus, being more stable and resistant to oxidation, Al is a highly desired cathode material. However, it has a higher work function. To reduce the injection barrier, interposing an ultrathin layer of LiF has stimulated a great deal of interest [8].

In some of the research papers [9-10], it has been reported that the LiF buffer layer at anode-organic interface and organic-organic interface improves the performance of the device. The purpose of this study is to verify these claims. In this paper we investigate the optimum position of LiF for enhancing the device performance.

EXPERIMENT

The multilayered green OLEDs were based on Aluminum (Al) as cathode, Lithium Fluoride (LiF) as buffer layer, Tris(8-hydroxyquinolinato)aluminium (Alq3) as electron transporting layer (ETL) and emissive layer (EL), *N,N′*-Bis(3-methylphenyl)-*N,N′*- diphenylbenzidine (TPD) as hole transporting layer (HTL), Poly(3,4-ethylenedioxythiophene):Poly(styrenesulfonate) (PEDOT: PSS) as anode buffer layer, and Indium Tin Oxide (ITO) as anode.

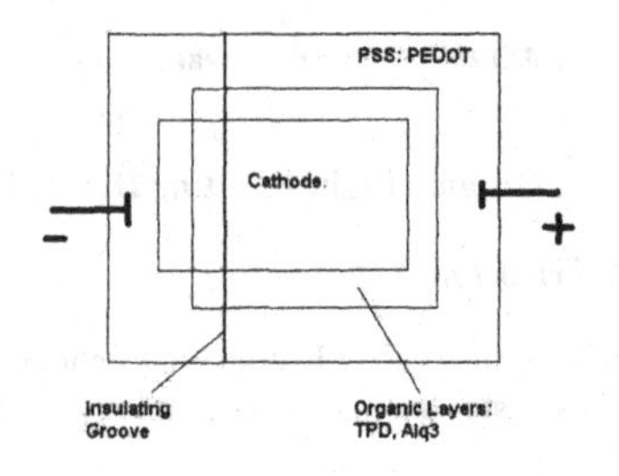

Figure 1. OLED device architecture

Figure 1 shows the device architecture. The OLEDs were fabricated by vacuum deposition and spin coating. ITO-coated glass with a sheet resistance of 30 ohm/□ was cut into a 1" × 1" plate and an insulating groove was created using a diamond cutter. The ITO substrate was then cleaned with 20% Ethanolamine at an elevated temperature, rinsed in DI water, and further cleaned by oxygen plasma treatment. PEDOT:PSS was spin coated on to the substrate. All the organic layres, LiF and Al were sequentially deposited under a pressure of 10^{-6} Torr. Thermal deposition rates for organic layers, LiF and Al were 5 A°/s, 0.1 A°/s and 1 A°/s, respectively. The active area of the devices was 0.25 cm^2. Luminance characteristics were measured using the Ocean Optics Spectrometer, while current-voltage measurements were performed using the Keithley 2400 Sourcemeter. L-I-V measurements were recorded simultaneously. All measurements were done in air at room temperature without any encapsulation. LiF thickness was varied between 0.5 to 2 nm. Best device characteristics were obtained with a 0.5 nm LiF thickness.

DISCUSSION

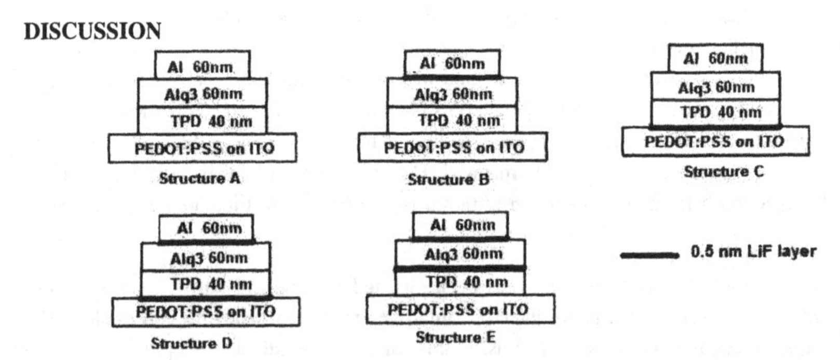

Figure 2. Structure A: No LiF; Structure B: LiF at Al/Alq3 interface; Structure C: LiF at Anode/TPD interface; Structure D: LiF at Al/Alq3 and Anode/TPD interfaces; Structure E: LiF at Al/Alq3 and TPD/Alq3 interfaces

Figure 2 shows a schematic of the devices fabricated for our study. Electroluminescent spectra show peak wavelength at 530 nm. Current-voltage characteristics of all the devices are plotted in

Figure 3 (a). Device B with LiF at cathode/Alq3 interface has steeper I-V characteristics with 20mA/cm^2 at 7.4 V. Device D, with LiF both at anode/TPD and cathode/Alq3 interfaces, shows even lower drive voltage with 20 mA/cm^2 at 6.7 V. Devices C and E do not show any improvement compared to the device without the LiF layer.

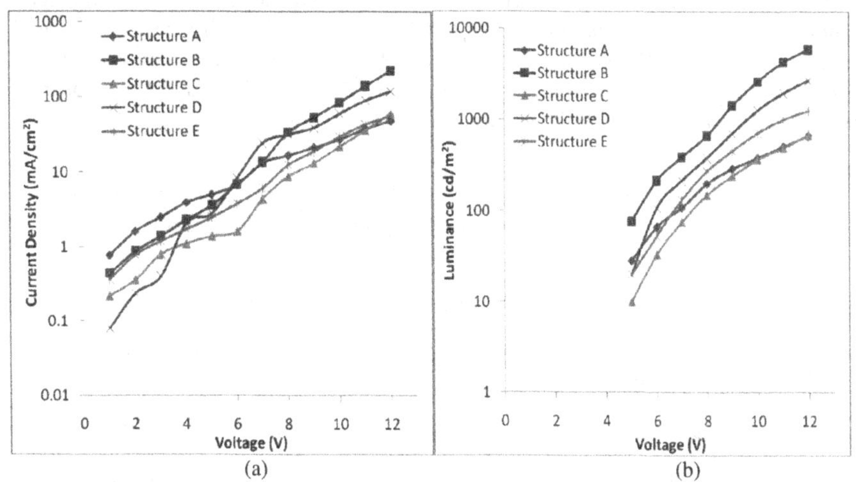

(a) (b)

Figure 3. (a) Current-Voltage Characteristics (b) Luminance characteristics

Luminance characteristics of all the devices are plotted in Figure 3 (b). Device B shows the highest luminance of 6000 cd/m2 at 12V. Luminance decreases with deposition of LiF at any other interface. Device C again shows no improvement in luminance compared to device A. Luminance yield is plotted in Figure 4 (a). Maximum luminance efficiencies of devices A to E are 1.42, 3.16, 2.04, 2.23 and 2.44 cd/A, respectively. Figure 4 (b) shows the energy level diagram.

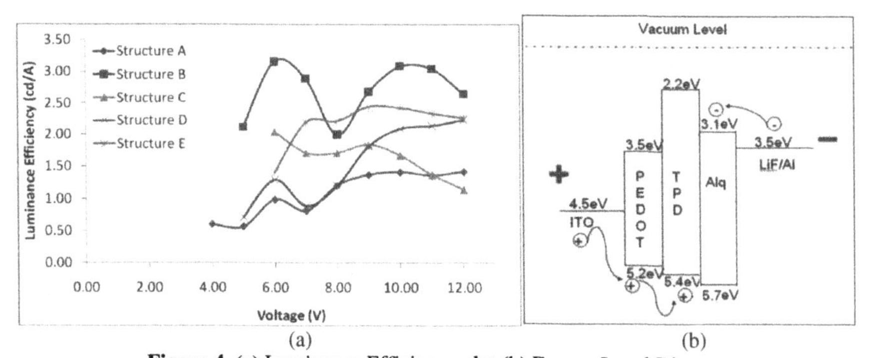

(a) (b)

Figure 4. (a) Luminance Efficiency plot (b) Energy Level Diagram

CONCLUSION

Inserting LiF at cathode/Alq3 interface clearly enhances the performance of the device. The device shows steeper I-V characteristics, improved luminance and luminance yield due to enhanced electron injection, carrier balance, and recombination efficiency. Device D, with LiF at both electrode interfaces, shows lowest drive voltage, possibly due to reduced hole injection shifting the recombination zone closer to Alq3/TPD interface. By comparing the performance of all the devices, this study rules out the possibility of improvement in the performance of the devices with structures D and E, as proposed in some of the research papers [9-10]. LiF is most effective only as a cathode buffer layer, enhancing OLED performance tremendously.

REFERENCES

1. R. H. Friend, R. W. Gymer, A. B. Holmes, J. H. Burroughes, R. N. Marks, C. Taliani, D. D. C. Bradley, D. A. Dos Santos, J. L. Bredas, M. Logdlund, and W. R. Salaneck, Nature (London) 397, 121 (1999).
2. J. C. Scott, S. Karg, and S. A. Carter, J. Appl. Phys., 82, 3, 1454–1460, (1997).
3. B. K. Crone, I. H. Campbell, P. S. Dabvids, D. L. Smith, C. J. Neef, andJ. P. Ferraris, J. Appl. Phys., 86, 10 (1999).
4. J. Kido, K. Nagai, and Y. Okamoto, IEEE Trans. Electron Devices 40, 1342 (1993).
5. T. Wakimoto, Y. Fukuda, K. Nagayama, A. Yokoi, H. Nakada, and M. Tsuchida, IEEE Trans. Electron Devices 44, 1245 (1997).
6. V. E. Choong, M. G. Mason, C.W. Tang, and Y. Gao, Appl. Phys. Lett. 72, 2689 (1998).
7. C. W. Tang and S. A. VanSlyke, Appl. Phys. Lett. 51, 913 (1987).
8. L. S. Hung, C. W. Tang, and M. G. Mason, Appl. Phys. Lett. 70, 152 (1997).
9. Y. Zhao, S.Y. Liu, J.Y. Hou, Thin Solid Films 397 (2001).
10. Xiao Jing, Deng Zhenbo, Liang Chunjun, Xu Denghui, Xu Ying, Guo Dong, Physica E 28 (2005).

Mater. Res. Soc. Symp. Proc. Vol. 1154 © 2009 Materials Research Society 1154-B05-87

Synthesis and Optical Properties of Perylene Bisimide Incorporated Low Bandgap Polymers for Photovoltaics

Sivamurugan Vajiravelu[1] and Suresh Valiyaveettil[1,*]

[1]S5-01-01, Materials Research Laboratory, Department of Chemistry, 3 Science Drive 3, National University of Singapore, Singapore 117543

ABSTRACT

Herein we report on the synthesis of perylene diimide (PDI) based P1 and P2 conjugated polymers via Suzuki polymerization. The chemical structure of the polymers was elucidated using GPC, ^{1}H, ^{13}C NMR and elemental analysis. The absorption spectra of polymers were in the visible region from 250 - 800 nm in solution and in solid state. The optical band gap was (E_g^{opt}) found to be between 1.60 – 1.83 eV in solid state.

INTRODUCTION

Synthesis and development of visible, IR and near IR absorbing materials are interesting to improve the solar cell efficiency [1]. Light harvesting devices based on perylene diimide (PDI) derivatives are well established and most attractive molecule possess high charge mobility, greater electron affinity and act as good *n*-type materials [2-5]. Enhanced intramolecular charge transfer has been achieved through covalent attachment of PDI as pendant or in main chain with different donor systems in a single molecule [6-9]. The polymer chain bearing alternative donor - acceptor moieties would alter the optical and electronic properties of mixed donor-acceptor system. This could improve stability and processability along with achieving stable energy transfer between donor-acceptor systems. Most of the polymers reported above are prepared via Yamamoto and Suzuki coupling reaction on the anhydride region of the perylene moiety. The bay region (1, 6, 7 and 12th position) of perylene diimide unit is a suitable position for functionalisation [10-14]. We report herein, PDI based conjugated polymer prepared using Suzuki polymerisation, in which donors such as fluorene and dithiophene are attached to 1st and 7th position of the perylene bisimide unit. The chemical structures of the polymers are shown in Figure 1.

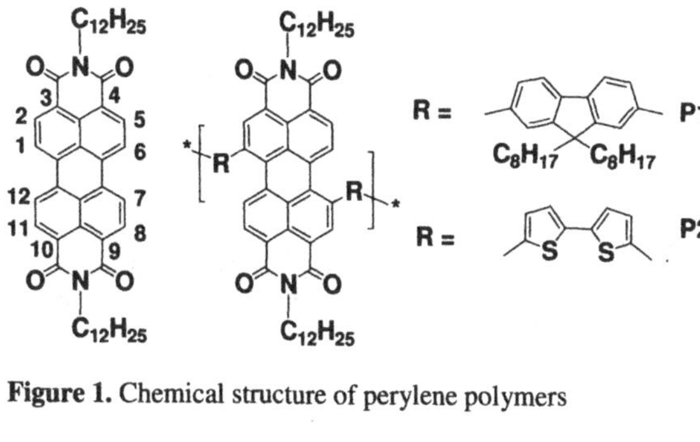

$C_{12}H_{25}$ O N O 3 4 2 5 1 6 12 7 11 8 10 9 O N O $C_{12}H_{25}$

$C_{12}H_{25}$ O N O * R R * O N O $C_{12}H_{25}$

R = (fluorene, C_8H_{17} C_8H_{17}) P1

R = (S S, bithiophene) P2

Figure 1. Chemical structure of perylene polymers

EXPERIMENT

Materials and methods

All chemicals and reagents were purchased from Aldrich and Acros, and used as such without further purification. The NMR spectral data of all synthesized samples were recorded on a Bruker ACF 300 spectrometer operating at 300 and 75 MHz for ^{1}H and ^{13}C nuclei, respectively, using $CDCl_3$ as solvent and TMS as internal standard. The absorbance and fluorescence spectra were recorded on a Shimadzu 1601 PC spectrophotometer and RF-5301PC Shimadzu spectrofluorophotometer, respectively. The phase behavior and thermal stability were identified by DSC and TGA using Universal V2.6D TA instruments.

Synthesis of polymers P1 and P2

1.12 mmol of PDI and 1.12 mmol of aryl diboronic acid were dissolved in 50 mL of THF (AR grade) and heated to 70 °C for 15 mins under nitrogen atmosphere. About 20 mL of 1M K_2CO_3 aqueous solution was added followed by addition of 10 mol% $Pd(PPh_3)_4$. The contents were heated at 75 °C under nitrogen atmosphere for 72 hrs. After completion, the reaction mixture was poured into excess methanol. The precipitated polymer was fitered and dried in hot air oven. The resulting polymer was purified by Schoxlet extraction using methanol and acetone.

P1: Reddish black solid. Yield 60 %. ^{1}H NMR (300 MHz, $CDCl_3$, δ ppm): 8.70 (s, 2H, Perylene-**H**), 7.40 – 8.17 (br, m, 11H, Per and Flu-**H**), 4.17 - 4.30 (br, s, 4H, **-CH2**-N), 2.0 – 2.30 (m, 5H, Flu-alkyl), 1.60 - 1.90 (m, 9H, alkyl-**CH_2-**), 1.32 - 1.42 (m, 38H, alkyl-**CH_2**), 0.60 – 1.20 (m, 44H, alkyl-**CH_3**). ^{13}C NMR (75 MHz, $CDCl_3$, δ ppm): 163.3 **(Per-C=O)**, 145.0, 141.6, 141.4, 135.1, 130.2, 129.4, 128.6, 128.3, 127.7, 122.4, 122.0, 78.4, 77.4, 77.2, 77.0, 76.58, 31.9, 29.6, 29.4, 29.3, 28.4, 28.1, 27.1, 22.6, 14.0, 13.9. Elem. Ana ($C_{77}H_{96}N_2O_4$) Calcd: C, 77.49; H, 8.11; N, 2.35. Found: C, 77.33; H, 8.12; N, 2.16.

P2: Black solid. Yield 74 %. ^{1}H NMR (300 MHz, $CDCl_3$, δ ppm) 8.65 - 8.62 (m, 4H, **Per-H**), 8.56 - 8.51 (m, 2H, **Per-H**), 8.36 - 8.28 (m, 4H, **Per-H**), 7.24 - 7.16 (m, 2H, **Th-H**), 7.10 - 7.02 (m, 2H, **Th-H**), 4.18 (m, 4H, **Per-N-CH_2**), 2.0 – 1.5 (m, 20H, **Per-N-alk**) 1.45 - 1.2 (m, 10H, **Per-N-alk**), 0.88 - 0.85 (m, 16H, **Per-alk**). ^{13}C NMR (75 MHz, $CDCl_3$, δ ppm): 162.3 (C=O), 163.2 (C=O), 163.1 (C=O), 163.1 (C=O), 134.4, 129.8, 129.7, 128.0, 127.9, 127.8, 125.3, 125.0, 124.5, 122.2, 77.4, 77.0, 76.57, 40.7 (**N-CH_2**), 31.9, 29.6, 29.5, 29.3, 29.3, 28.1, 27.1, 22.6, 14.0. Elem. Ana ($C_{56}H_{60}N_2S_2O_4$) Calcd: C, 69.40; H, 6.24; N, 2.89. Found: C, 69.92; H, 6.22; N, 2.72; S, 6.01.

DISCUSSION

Synthesis of polymers

The polymers P1 and P2 have been prepared according to Scheme 1. The comonomers such as fluorene and dithiophene diboronic acids coupled to 1st and 7th position of N,N'-didodecyl-1,7-dibromoperylene bisimide, using Suzuki coupling technique [10-14]. The monomer, PDI was prepared by controlled bromination of perylene dianhydride using Br_2/H_2SO_4 in presence of I_2 as catalyst which afforded 80% of 1,7-dibromo perylene. Further, the imide moiety was introduced using n-dodecyl amine in N-methyl pyrrolidone medium in the presence of acetic acid which afforded 75% of PDI. The structure of PDI was confirmed using ^{1}H and ^{13}C NMR and MS techniques (Scheme 1). The PDI polymers were prepared by Suzuki polymerisation (Scheme 1). After 3 days of stirring, the polymer was preciptated from 50% methanol solution. The polymers were purified by Soxchlet extraction using methanol and acetone solvent. P1 and P2 were obtained in 65 and75 % yield, respectively.

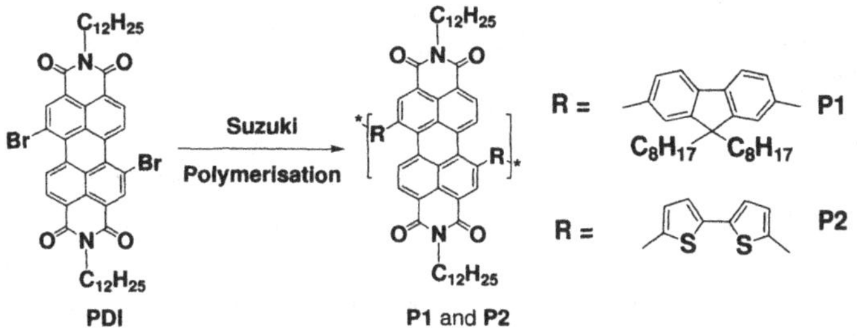

Scheme 1. Synthesis of polymers

Molecular weight and physical properties of the polymers

GPC analysis was performed on a Shimadzu LC10AT instrument using THF as eluent. The molar masses of the polymers were determined using polystyrene standards. **Table 1** shows the molecular weight and poly dispersity index (PDI) of the polymers.

Table 1. Physical data of the polymers **P1**and **P2**

Polymer	Mn (kD)	Mw (kD)	D	Tg (°C)	Td (°C)
P1	11.0	16.0	1.45	75	412
P2	12.4	17.3	1.39	101	495

Absorption spectra

The absorption spectra of polymers and precursors were recorded using normalised concentration of 1 mg/25 mL in chloroform and in thin films (Figure 2). The optical bandgap are found to be 1.91, 1.69 and 2.21 eV for P1, P2 and PDI, respectively. PDI showed two absorption maximum at 489 and 525 nm corresponding to π - π^* transition. The perylene polymers showed wide range of absorption band from 300 to 700 nm, which covers the whole range of visible region. In particular, fluorene attached perylene polymer (P1) showed three absorption maxima at 336, 449 and 544 nm, which correspond to fluorene moiety, perylene π - π^* transition and donor-acceptor charge transfer band, respectively. The bathochromic shift and disappearence of vibronic fine structure of absorption band may be due to twisting of perylene core at the bay region [12-15]. Dithiophene substituted polymer, P2 showed broad absorption in the range of 300 – 750 nm with absorption maxima at 600 nm. The P1 has higher bandgap as compared to P2 due to the rigidity and bulkiness of fluorene ring. So the nature of spacer attached between two perylene units and donor ability play a crucial role in the optical properties.

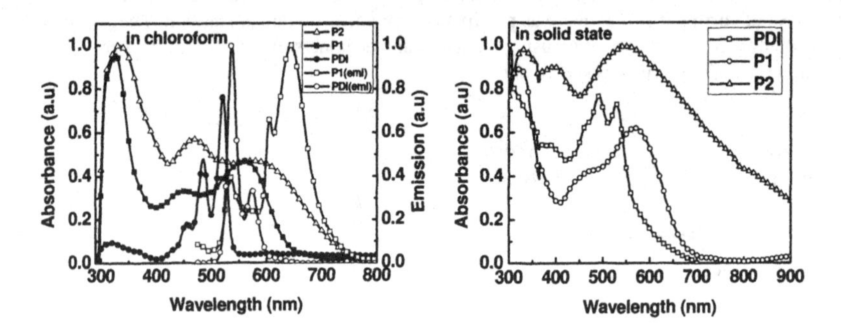

Figure 2. Absorption and emission spectra of P1, P2 and PDI in chloroform solution and spectra in thin films

Fluorescence studies

Emission spectra of the polymers and the monomer are shown in Figure 2. The monomer showed two sharp and well separated emission peaks at 540 and 570 nm. The polymer, P1 showed emission maximum with high intensity at 640 nm when excited at 335, 440 and 545 nm. The emission from fluorene attached perylene polymer, P1 is red in colour whereas PDI showed an yellow emission. No observable emission was found from polymer P2 and it may be due to self-quenching of fluorescence [14-16]. The emission properties of the perylene polymers can be tuned by attaching with suitable donor units in the bay region.

Thermal stability and T_g of the polymers

The thermal stability and T_d of the polymers were obtained by TGA (Model: 2960 SDT V3.0F, Processing software: Universal V3.9A TA instruments) and T_g was identified using DSC. TGA and DSC curves of PDI, P1 and P2 are shown in Figure 3. TGA traces were recorded using double beam TG analyser (TA instruments) with nitrogen as carrier gas at a heating rate of 20 °C. min^{-1} and temperature range from 30 to 900 °C (Figure 3a). The polymers were thermally stable upto 350 - 450 °C. The 5 wt% of initial weight loss which observed between 200 and 300 °C may be due to removal of residual solvents. The major 15 wt% second weight loss gradually observed after 400 °C may be due to degradation of dodecyl chain. Fluorene group attached perylene polymer, P1 showed very high weightloss (35 wt%) compared to P2 due to the presence two 2-ethylhexyl side chains attached in fluorene moiety. The third major weight loss observed between 500 – 550 °C is due to the removal of carbonyl group present in the perylene unit as carbon dioxide. After 550 °C, no major weight loss was observed upto 900 °C. Figure 3b showed DSC curves of PDI, P1 and P2. PDI showed sharp melting at 155 °C without any glass transition temperature. P1 and P2 showed T_g at 75 and 101 °C, respectively. The DSC results revealed that polymers were in amorphous nature.

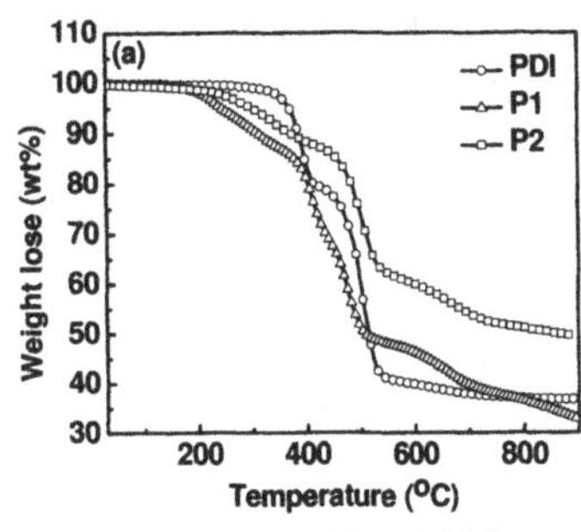

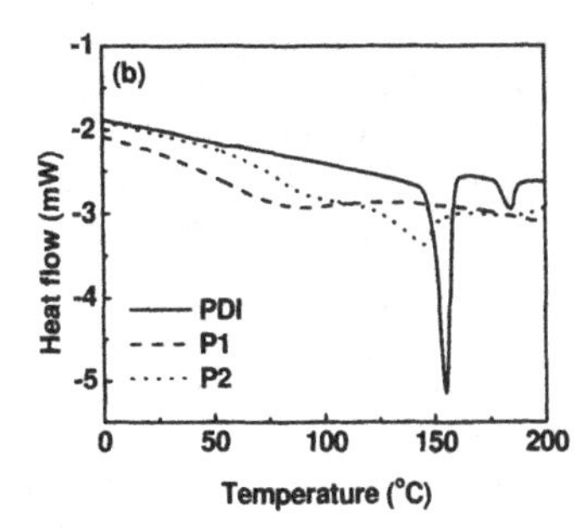

Figure 3. TGA traces (a) and DSC traces (b) of polymers recorded using nitrogen as a carrier gas at a heating rate 20 °C/min.

CONCLUSIONS

In summary, perylene bisimide based polymers, P1 and P2 with electron donating substituents have been synthesized in good yield and characterized. The absorption and emission spectra revealed considerable bathochromic shift compared to unsubstituted PDI. Broad absorption from these compounds suggests an extended π - π conjugation due to the presence electron donating aryl groups in the bay region of PDI and expected intramolecular energy transfer. The presence of electron donating aryl units lowered the band gap significantly.

ACKNOWLEDGMENTS

The financial support received from Agency for Science, Technology and Research (A*STAR) Singapore is gratefully acknowledged.

REFERENCES

1. S. Gunes, H. Neuebauer and N. S. Sariciftci, *Chem. Rev.* **107,** 1324 (2007).
2. J. Y. Do, B. G. Kim, J. Y. Kwon, W. S. Shin, S. H. Jin and Y. I. Kim, *Macromol. Symp.* **249-250,** 461 (2007).
3. Y. Li, Y. Liu, N. Wang, Y. Li, H. Liu, F. Lu, J. Zhuang and D. Zhu, *Carbon.* **43,** 1968 (2005).
4. A. Burquel, V. Lemaur, D. Belijonne, R. Lazzaroni and J. Cornil, *J. Phys. Chem. A.* **110,** 3447 (2006).
5. I. Yutaka, U. Toshihiko, Y. Nobuhiro and A. Yoshio, *Chem. Commun.* 1213 (2009).
6. Y. Liu, C. Yang, Y. Li, Y. Li, S. Wang, J. Zhuang, H. Liu, N. Wang, X. He, Y. Li and D. Zhu, *Macromolecules,* **38,** 716(2005).

7. Z. Zhu, D. Waller, R. Gaudiana, M. Morana, D. Muhlbacher, M. Scharber and C. Brabec, *Macromolecules,* **40,** 1981 (2007).
8. R. Yang, R. Tian, J. Yan, Y. Zhang, J. Yang, Q. Hou, W. Yang, C. Zhang and Y. Cao, *Macromolecules,* **38,** 244 (2005).
9. K. Colladet, S. Fourier, T. J. Cleij, L. Lutsen, J. Gelean, D. Vanderzande, L. H. Naguyen. H. Neugebauer, S. Sariciftci, A. Aguirre, G. Janssen and E. Goovaerts, *Macromolecules,* **40,** 65 (2007).
10. X. He, H. Liu, N. Wang, X. Ai, S. Wang, Y. Li, C. Huang, S. Cui, Y. Li and D. Zhu, *Macromol. Rap. Commun.* **26,** 721 (2005).
11. X. Zhan, Z. Tan, B. Domercq, Z. An, X. Zhang, S. Barlow, Y. Li, D. Zhu, B. Kippelen and S. R. Marder, *J. Am. Chem. Soc.* **129,** 7246 (2007).
12. J. Hou, S. Zhang, T. Chen and Y. Yang, *Chem. Commun.* 6034 (2008).
13. L. Hou, Y. Zhou and Y. Li, *Macromol. Rapid Commun.* 1444 (2008).
14. V. Sivamurugan, R. Lygaitis, J. V. Grazulevicius, V. Gaidelis, V. Jankauskas and S. Valiyaveettil, *J. Mater. Chem.* (2009) (DOI: 10.1039/B901847F).
15. M. J. Ahrens, M. J. Tauber and M. R. Wasielewski, *J. Org. Chem.* **71,** 2107 (2006).
16. S. Chen,Y. Liu, W. Qiu, X. Sun, Y. Ma and D. Zhu, *Chem. Mater.* **17,** 2208 (2005).

Modification of Semiconductor-Dielectric Interface in Organic Light-Emitting Field-Effect Transistors

Yan Wang,[1] Ryotaro Kumashiro,[1] Naoya Komatsu,[1] and Katsumi Tanigaki[1,2]
[1]Department of Physics, Graduate School of Science, Tohoku University, Sendai, 980-8578, Japan
[2]World Premier International Research Center, Tohoku University, Sendai, 980-8578, Japan

ABSTRACT

In this work, ambipolar rubrene single crystal field-effect transistors (FETs) with PMMA modification layer and Au/Ca as electrodes were fabricated. The electron mobility was studied in these devices. PMMA modification layer on the surface of SiO_2 is necessary for electron behavior. We found that the device with PMMA modified insulator and Au-Ca asymmetric metals possessed hole mobility and electron mobility of 1.27 and 0.017 cm^{-2}/V s, respectively. Furthermore, the shift of light emitting with applied gate voltage was observed in this device.

INTRODUCTION

Light-emitting field-effect transistors (LE-FETs) are attracting much attention due to their applications in lighting, displays and circuits [1-3]. Light-emitting FETs on Si-based transistors are difficult to be achieved because of the very weak light emitting probability from the excitonic states in silicon due to its inherent indirect band gap; while organic semiconductors are known to be able to show very efficient electro-luminescence. In addition, ambipolar characteristics are possible in organic semiconductors, i.e., both electrons and holes can simultaneously be injected into a uniform semiconducting layer to create excitonic states, which enables strong light emission [4]. These recent results have triggered scientific community to have further intensive studies on exploring organic light-emitting field-effect transistors (OLE-FETs). However, the charge carriers, especial electrons, are easily trapped when they are injected from electrodes or during their transportation and accumulation in the interface between the semiconductor and the dielectric layer [5, 6]. The inefficient electron transportation in the interfaces and injection from electrodes result in the low electro-luminescence quantum efficiency. In order to improve the electro-luminescence quantum efficiency, many researches are being carried out to increase the injection efficiency of electron from electrode as well as the transport efficiency in the channel. In this work, we studied the electron mobility in the FET devices with PMMA modification layer and Au/Ca as electrodes. Also, we chose the rubrene single crystals as the organic semiconductor due to its high hole mobility [7].

The rubrene single crystals were grown by physical vapor deposition in a stream of argon gas [8]. A heavily doped silicon wafer with a 200-nm thermally grown SiO_2 layer (ε= 3.9) was used as the substrate. The 100-nm polymethylmethacrylate (PMMA) film (ε= 3.7) was spin coated onto the SiO_2 surface. Rubrene single crystals were laminated on these substrates by electrostatic interaction. The top contact electrodes were deposited by thermally evaporating Au and Ca metals through a shadow mask on top of the rubrene single crystals [9, 10]. The length of the channel is 50 μ m in all the devices and the width of the channel varied with the single crystal. The structure of FETs is shown in figure 1. Electrical characterizations were performed inside a glove box in argon gas by using a semiconductor device analyzer (Agilent Technologies B150A). Light emitting images were recorded from a CCD camera.

DISCUSSION

For rubrene single crystal, its HOMO level (5.36 eV) is close to the Fermi level of Au (5.1 eV) as can be seen in Figure 1, so the hole carriers are favored to be injected in FETs with Au-Au electrodes, while electron injection would be rather difficult. However, a high efficient electron carrier injection is crucial to utilize ambipolar properties for the realization of high efficient light emitting transistors. So far, many reports suggested that low work function electrode could give better electron carrier injection to the conduction band. In contrast to the Au electrodes, the work function of Ca is 2.87 eV, which matches the LUMO level of rubrene much better. So it is easily understood that the injection efficiency of electron from Ca electrode should be higher than which from Au electrode.

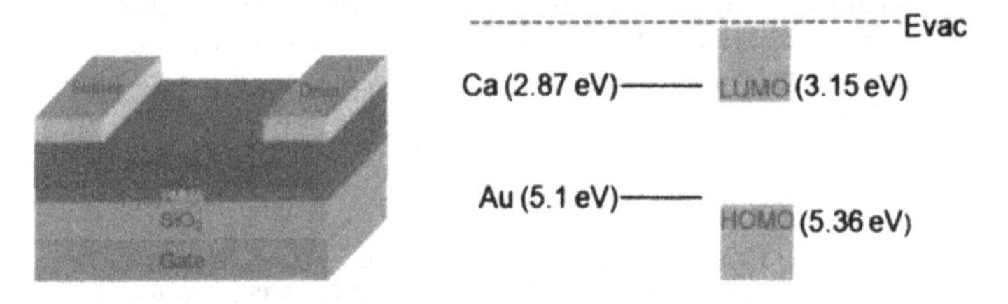

Figure 1. Left: The structure of field-effect transistor. Right: Highest occupied molecular orbital (HOMO) and lowest unoccupied molecular orbital (LUMO) levels of rubrene in relation to the work functions of Ca and Au.

Therefore, we made a device with Au-Ca asymmetric electrodes on the untreated SiO_2. The transfer characteristic of this device is given in figure 2a. Unfortunately, we could not observe any electron current from this device, though the electron injection efficiency from Ca is high. This is because all the injected electrons were trapped when then transported in the channel near the surface of SiO_2. Chua et al. confirmed that the injected electron from electrode were completely trapped at the interface between the semiconductor and insulator due to the electrochemical reaction and generated SiO^- groups [5]. Hence, it is necessary to modify the surface of SiO_2 in order to decrease or eliminate the electron trapping. A possible modification method is using non-hydroxyl polymer, for example, PMMA, which has the ability to terminate the electron traps on the surface of SiO_2 [4]. Therefore, the PMMA modification layer was spin coated onto the substrate in order to decrease the electron traps on the surface of SiO_2. The electron mobilities were investigated in the following devices with PMMA modification layer as well as with Au-Au, Ca-Ca, and Au-Ca as electrodes.

Figure 2b shows the transfer characteristic of device with PMMA modification layer and Au-Au symmetric electrodes. One can see that though the energy mismatch between the Au electrode and the LUMO level of rubrene is too large for electron injection by thermal energy alone, it still behaved as the electron injection from Au electrode. We estimated the electron mobility according to the equation of

$$I_{DS,Sat} = \frac{W}{2L} \mu_{FET} C_{in} (V_G - V_{th})^2$$

and the value of 6.7×10^{-4} cm^{-2}/V s was obtained. The reason for electron behavior in this device is the low electron traps density in the interface between the rubrene single crystal and insulator due to the modification of interface between semiconductor-dielectric. Meanwhile, the trap states at the Au/rubrene interface introduced by the thermal evaporation of Au also assistant the electron injection, as reported by Takenobu et al [11]. We also estimated the hole mobility in this device in the p-channel operation and the value is 0.47 cm^{-2}/V s. Figure 2c shows the transfer characteristic of the device with PMMA modification layer and Ca-Ca symmetric electrodes in n-channel operation. The extracted electron mobility in this device was 0.11 cm^{-2}/V s. It is about three orders higher than that from Au. This is because the injection of electron from Ca electrode is much higher than which from Au electrode. As it can be seen, the holes should suffer from the large injection barrier caused by the difference between Ca work function and the HOMO level of rubrene single crystal. However, the hole current can still be observed in the transfer characteristic. One possibility is that the holes are injected via in-gap states at the interface between Ca electrode and the rubrene single crystals. These in-gap states, with various energy levels, are scattered in between the HOMO and LUMO levels, which are induced by the defects created at rubrene surface beneath the Ca electrode due to thermal evaporation. The holes can hop from one to another in-gap state before arriving at the LUMO level of rubrene single crystal [11]. From these results, one can see that in rubrene single crystal FETs, the high hole mobility can been obtained by using Au as electrode and high electron mobility can been obtained by using Ca as electrode with PMMA modification layer on the surface of SiO_2. Therefore, the remarkably ambipolar behavior could be shown in the device with PMMA modification layer and Au-Ca asymmetric electrodes. The transfer characteristic of this device is given in figure 2d. The performance of its ambipolar behavior is better in this device with estimated hole mobility of 1.27 cm^{-2}/V s and electron mobility of 0.017 cm^{-2}/V s. Furthermore, we noticed that the electron and hole mobilities are rather close, which show potential applications in light emission devices. Meanwhile, the electron mobility is a little lower compared with that obtained in the device with Ca-Ca symmetric configuration. According to the other report, this problem occurs possibly due to the difference from the contact size area between the symmetric device and asymmetric device (see figure 3), especially because the sandwich structure of both different metal [12, 13].

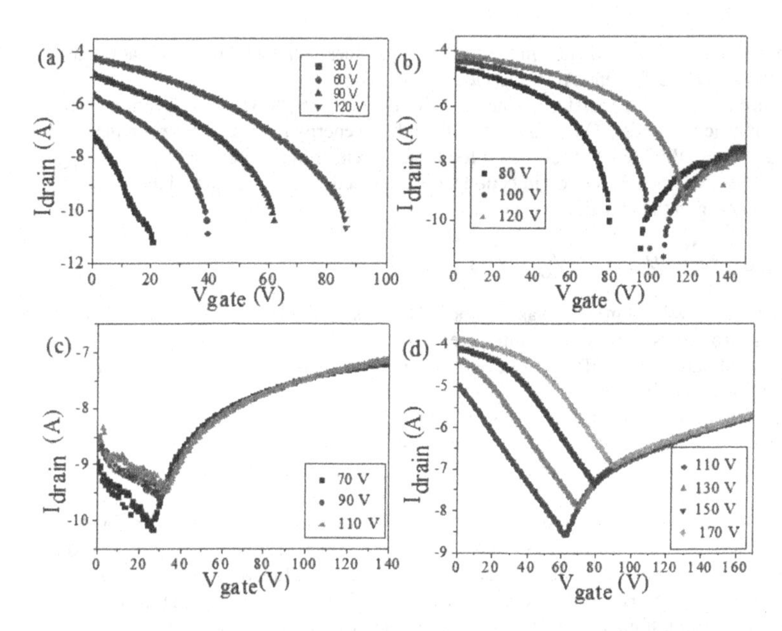

Figure 2. Transfer characteristics of rubrene single crystal FETs with (a) Au-Ca asymmetric electrodes and without PMMA modification layer. (b) Au-Au symmetric electrodes and with PMMA modification layer. (c) Ca-Ca symmetric electrodes and with PMMA modification layer. (d) Au-Ca asymmetric electrodes and with PMMA modification layer.

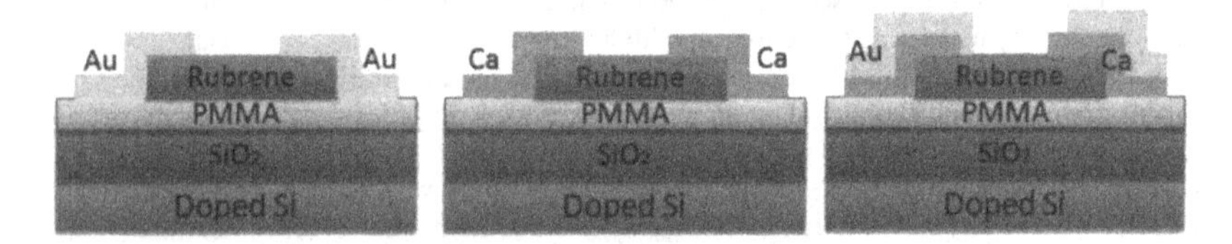

Figure 3. The structure of FETs with symmetric or asymmetric electrodes.

In our study, a high electron mobility was obtained from the device with PMMA modification layer and Au-Ca asymmetric electrodes, which is crucial for the light emitting organic FETs. Figure 4 shows a series of CCD images with various V_G of the rubrene single crystal transistor with Au-Ca asymmetric and PMMA modified insulator at fixed V_D (150 V). Below a certain V_G, no ambipolar transportation or light emission could be observed as the channel was predominant occupied only by hole. When V_G increased, the relative bias between source and gate exceeded the threshold voltage for the formation of an electron accumulation layer. Very clear light emitting appeared alongside the edge of the electron-injecting Ca electrode. And the emission zone shifted through the channel towards the Au electrode with increasing V_G as the electron accumulation region extended further into the channel. Finally, the light position was fixed close to the Au electrode with continuous increasing intensity with the increase of gate voltage.

Figure 4. Shift of light emitting from a rubrene single crystal transistor with Au-Ca asymmetric electrode at fixed V_D (V_D = 150 V) and different V_G.

CONCLUSIONS

In conclusion, the electron mobility in the devices with Au and Ca as electrodes was studied systematically. In our research, the modification of semiconductor-dielectric interface is necessary for electron behavior. The high electron mobility was obtained from the devices with PMMA modification layer and Au-Ca asymmetric electrodes due to the high electron injection efficiency from Ca into the LUMO level of rubrene single crystal and low electron trap density in the interface between rubrene single crystal and SiO_2. The nearly balanced hole mobility of 1.27 cm^{-2}/V s and electron mobility of 0.017 cm^{-2}/V s show that this device has a high performance. Furthermore, a bright luminescence was observed in the device with PMMA modification layer and Au-Ca asymmetric electrodes. The shift of light emitting was observed with applied gate voltage, which means that we can tune the recombination/emission zone to any position in the channel.

ACKNOWLEDGMENTS

This work was supported by Grants-in-Aids (No.18204030, 19014001, 18651075 and 18204032) and for Scientific Research on Priority Areas "(New Materials Science Using Regulated Nano Spaces-Strategy in Ubiquitous Elements") from the Ministry of Education, Culture, Sports, Science and Technology of Japan. The research was also partially supported by Tohoku University GCOE program. Y. Wang also thanks the Japan Society for the Promotion of Science (JSPS) for supporting of Scientific Research (Grant No. P08372).

REFERENCES

1. H. Sirringhaus, *Adv. Mater.* **17,** 2411 (2005).
2. B. W. D'Andrade, and S. R. Forrest, *Adv. Mater.* **16,** 1585 (2004).
3. M. L. Chabinyc, and A. Salleo, *Chem. Mater.* **16,** 4509 (2004).
4. T. Takenobu, S. Z. Bisri, T. Takahashi, M. Yahiro, C.Adachi, and Y. Iwasa, *Phys. Rev. Lett.* **100,** 066601 (2008).

5. L. L. Chua, J. Zaumseil, J. Chang, E. C.-W. Ou, . K.-H. Ho, H. Sirringhaus, and R. H. Friend, *Nature*, **434,** 194 (2005).
6. C. R. Newman, C. D. Frisbie, D. A. da S. Filho, J. Bredas, P. C. Ewbank, and K. R. Mann, *Chem. Mater.* **16,** 4436 (2004).
7. E. Menard, V. Podzorov, S.-H. Hur, A. Gaur, M. E. Gershenson, and J. A. Rogers, *Adv. Mater.* **16,** 2097 (2004).
8. C. Reese, and Z. Bao, *Materials Today,* **10,** 20 (2007).
9. J. Takeya, C. Goldmann, S. Haas, K. P. Pernstich, B. Ketterer, and B. Batlogg, *J. Appl. Phys.* **94,** 5800 (2003).
10. J. Takeya, T. Nishikawa, T. Takenobu, S. Kobayashi, Y. Iwasa, T. Mitani, C. Goldmann, C. Krellner, B. Batlogg, *Appl. Phys. Lett.* **85,** 5078 (2004).
11. T. Takenobu, T. Takahashi, J. Takeya, and Y. Iwasa, *Appl. Phys. Lett.* **90,** 013507 (2007).
12. S. Z. Bisri, T. Takahashi, T. Takenobu, M. Yahiro, C. Adachi, and Y. Iwasa, *Jpn. J. Appl. Phys.* **46,** L596 (2007).
13. S. Z. Bisri, T. Takenobua, Y. Yomogidaa, T. Yamaoc, M. Yahirod, S. Hottac, C. Adachid, and Y. Iwasa, *Proc. of SPIE* **6999,** 69990Z (2008).

Mater. Res. Soc. Symp. Proc. Vol. 1154 © 2009 Materials Research Society 1154-B05-98

Near Infrared Fluorescent and Phosphorescent Organic Light-Emitting Devices

Yixing Yang,[1] Richard T. Farley,[2] Timothy T. Steckler,[2] Jonathan Sommer,[2] Sang-Hyun Eom,[1] Kenneth R. Graham,[2] John R. Reynolds,[2] Kirk S. Schanze,[2] and Jiangeng Xue[1]
[1] Department of Materials Science and Engineering, University of Florida, Gainesville, Florida 32611, USA
[2] Department of Chemistry, Center for Macromolecular Science and Engineering, University of Florida, Gainesville, Florida 32611, USA

ABSTRACT

Organic light-emitting devices (OLEDs) emitting near-infrared (NIR) light have many potential applications, yet the efficiency of these devices remains very low, typically ~0.1% or less. Here we report efficiency NIR OLEDs based on two fluorescent donor-acceptor-donor oligomers and a phosphorescent Pt-containing organometallic complex. External quantum efficiencies in the range of 0.5-3.8% with emission peak ranging from 700 to 890 nm have been achieved.

INTRODUCTION

There has been a growing interest in the development of near-infrared (NIR) organic light-emitting devices (OLEDs) due to their potential applications in security and defense, biomedical devices, and telecommunications [1-3]. Existing NIR OLEDs are primarily based on lanthanide complexes [1,4-5], but they generally have very low external quantum efficiencies (EQE~0.1%) because of the inherent low efficiency of emission from these complexes. Only recently were high efficiency NIR OLEDs reported, in which peak emission at $\lambda \approx 770$ nm and a maximum EQE up to 8.5% were achieved using a phosphorescent Pt-porphyrin complex [6]. Nonetheless, alternative materials and devices that exhibit longer wavelength ($\lambda > 800$ nm) emissions with high efficiencies are still needed.

π-conjugated donor-acceptor-donor (DAD) oligomers and polymers have been extensively investigated for their tunable optoelectronic properties [7]. In these DAD molecules, it is possible to tune their energy gap by using appropriate structures of the donor and acceptor units, which control the energies of the highest occupied molecular orbital (HOMO) and the lowest unoccupied molecular orbital (LUMO), respectively. By covalently combining strong donors with strong acceptors, a set of DAD oligomers and polymers with narrow HOMO-LUMO gaps have been achieved [8-9]. These oligomers have especially long wavelength absorption and efficient photoluminescence (PL), making them excellent candidates as NIR light emitting species in OLEDs.

Alternatively, a family of complexes of metalloporphyrins has shown intense absorption and emission in the red-to-NIR region. High efficiencies can be achieved when they are used as the phosphorescent emitters in OLEDs [6]. By extending the conjugation of the porphyrin core, the emission can be further shifted to longer wavelengths.

Here, we report NIR OLEDs based on two DAD conjugated oligomers, 4,8-bis(2,3-dihydrothieno-[3,4-*b*][1,4]dioxin-5-yl)benzo[1,2-*c*;4,5-*c*′]bis[1,2,5]thiadiazole

(BEDOT-BBT) and 4,9-bis(2,3-dihydrothieno[3,4-*b*][1,4]dioxin-5-yl)-6,7-dimethyl-[1,2,5] thiadiazolo[3,4-*g*]quinoxaline (BEDOT-TQMe$_2$). Using tris(8-hydroxyquinoline) aluminum (Alq$_3$) as the host in the emissive layer, we have achieved a maximum EQE of 1.6% in the device based on BEDOT-TQMe$_2$. The electroluminescence (EL) of this device peaks at λ = 692 nm, and extends to well above 800 nm. With a stronger electron acceptor in the molecule, the BEDOT-BBT based OLED had a peak emission at λ = 815 nm, extending to as far as 950 nm. A maximum EQE of 0.5% is achieved in this latter device. By incorporating a phosphorescent sensitizer in the emissive layer to funnel the triplet excitons formed on the host molecules to the fluorescent emitters, which is so-called "sensitized fluorescence" [10], we achieved maximum efficiencies of η_{EQE} = 3.1% and η_P= 12 mW/W for BEDOT-TQMe$_2$ based devices, and η_{EQE} = 1.5% and η_P= 4.0 mW/W for BEDOT-BBT based devices.

Furthermore, NIR OLEDs based on a metalloporphyrin, platinum tetraphenyltetranaphtho[2,3]porphyrin (Pt-TPTNP), have also demonstrated. Peak emissions at 892 nm and maximum efficiencies of η_{EQE} = 3.8% and η_P= 19 mW/W have been achieved. This is the most efficient NIR OLED reported to date for emissions beyond 800 nm.

EXPERIMENTAL

The molecular structures of two DAD oligomers (BEDOT-TQMe$_2$ and BEDOT-BBT) and Pt-TPTNP are shown in Figure 1. The synthesis of these molecules has been reported previously [8]. The measurement for UV-Vis absorption and photoluminescence (PL) spectra were followed by the methods described in Ref. 11.

OLEDs are fabricated on indium-tin-oxide (ITO) coated glass substrates, and all organic layers and the cathode were deposited using the vacuum thermal evaporation method in a high vacuum chamber with a base pressure of ~10^{-7} Torr. To fabricate a fluorescent OLED, the hole transport layer (HTL) of bis[N-(1-naphthyl)-N-phenyl-amino]biphenyl (α-NPD), the emissive layer (EML) consisting of Alq$_3$ doped with either fluorescent NIR emitter, and the electron transport layer (ETL) of bathocuproine (BCP) were successively deposited on the ITO anode, which was cleaned in solvent and treated in a UV-ozone environment. While the thickness of the α-NPD and Alq$_3$ layers were maintained at 40 nm and 20 nm, respectively, the doping concentration of NIR emitters in the EML and BCP layer thickness was optimized to achieve the highest device efficiencies. A 1 nm thick layer of LiF followed by a 50 nm thick Al layer was then deposited as the cathode.

For the sensitized fluorescent devices, 4,4′-bis(carbazol-9-yl)biphenyl (CBP) was used as the host material in EML, which also incorporates a phosphorescent sensitizer in addition to the NIR emitters. tris(2-phenylpyridine)iridium(III) (Ir(ppy)$_3$) and bis(2-phenylquinoline) (acetylacetonate)iridium(III) (PQIr) were used as the phosphorescent sensitizers, both with 10 wt.% doping concentration, for BEDOT-TQMe$_2$ and BEDOT-BBT devices, respectively.

For the phosphorescent devices, the EML consisted of CBP doped with 8 wt.% Pt-TPTNP. Instead of BCP, 4,7-diphenyl-1,10-phenanthroline (BPhen) was used as the ETL with optimized thickness of 100 nm.

Radiant emittance (R) – current density (J) – voltage (V) characteristics and EL spectra of the NIR OLEDs were also measured following the approaches in Ref. 11. The EQE (η_{EQE}) and power efficiency (η_P) were derived based on the recommended methods [12].

RESULTS AND DISCUSSION

The absorption and PL spectra of the two DAD molecules (in CH_2Cl_2) and Pt-TPTNP (in toluene) are shown in Figure 1. While both DAD molecules show strong absorption at λ = 350 to 370 nm, BEDOT-TQMe$_2$ has a longer wavelength absorption peak at λ = 531 nm, which is significantly redshifted to λ = 650 nm for BEDOT-BBT, due to the stronger electron acceptor component in BEDOT-BBT, leading to a lower lying LUMO level and thus a lower energy gap. Correspondingly, the PL of BEDOT-BBT, peaked at λ = 805 nm, is also significantly red shifted from that of BEDOT-TQMe$_2$, which has a peak at λ = 698 nm. The PL quantum yield for BEDOT-TQMe$_2$ is measured to be 21% while that for BEDOT-BBT is approximately three times lower, only 7.6%. For the phosphorescent Pt-TPTNP molecule, it exhibits Soret absorption peaked at λ = 436 nm and a strong absorption peak at λ = 689 nm. The PL of the complex has a phosphorescence emission peaked at λ = 883 nm with the quantum yield of 22%.

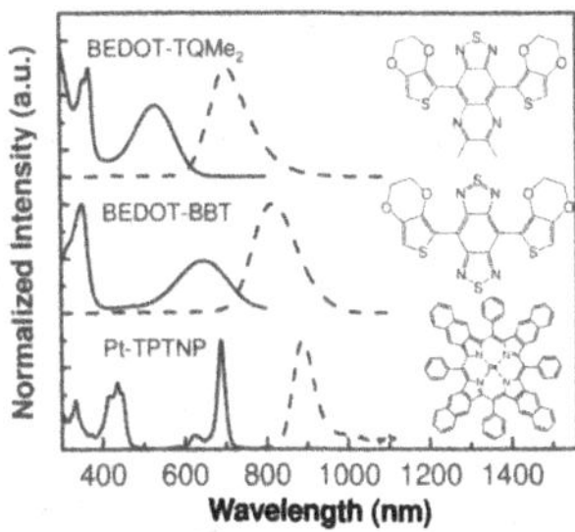

Figure 1. Absorption (solid lines) and PL (dash lines) spectra of two DAD oligomers (in CH_2Cl_2) and Pt-TPTNP (in toluene). The structure of these molecules is also shown.

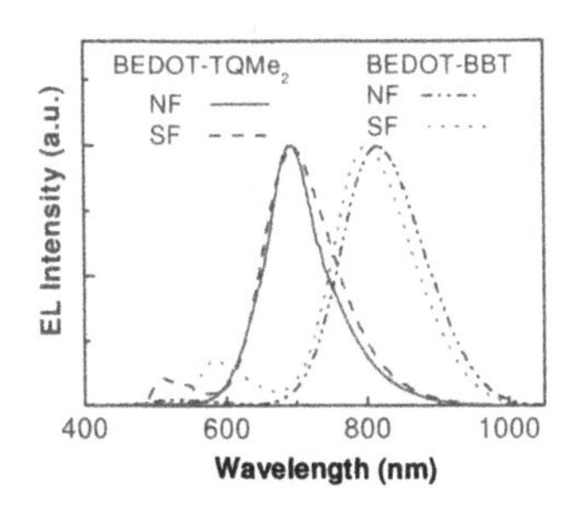

Figure 2. Normalized EL spectra, for (normal) fluorescent (NF) and sensitized fluorescent (SF) OLEDs based on BEDOT-TQMe$_2$ and BEDOT-BBT.

In the fluorescent devices based on these two DAD oligomers, the optimized doping concentration of the oligomer in Alq_3 host is 3.5% for both molecules, to achieve a proper balance between the maximum energy transfer and minimum concentration quenching. The optimal BCP ETL thicknesses are found to be 80 nm and 100 nm for devices based on BEDOT-TQMe$_2$ and BEDOT-BBT, respectively, to maximize the coupling of the emission into the external modes. For the sensitized fluorescent devices, BEDOT-TQMe$_2$ doping concentration is slightly decreased from 3.5% to 3% whereas that of BEDOT-BBT is maintained at 3.5%. The doping concentration of both phosphorescent sensitizers, $Ir(ppy)_3$ and PQIr, is 10 wt.%. The ETL thicknesses are the same as in the optimized normal fluorescent devices.

Figure 2 shows the EL spectra of the NIR OLEDs based on the two DAD molecules, measured at J = 1 mA/cm^2. Here, "NF" stands for (normal) fluorescence and "SF" for sensitized fluorescence. For the NF devices, with a 3.5% doping concentration of BEDOT-TQMe$_2$ in the EML, the Alq_3 host emission, which should appear at around λ = 520 nm, is completely quenched and only emission from dopant molecules, which peaks at λ = 692 nm, is observed. For the BEDOT-BBT based NF device, the EL spectrum shows a prominent NIR emission peak at λ = 815 nm while the host emission is barely apparent with an intensity of approximately only 3% of that of the NIR peak. Both these fluorescent OLEDs show relatively broad emissions with full widths at half maximum (FWHM) of approximately 120 nm.

The emission peak wavelengths of the SF devices are very close to the corresponding NF devices. However, unlike the NF devices, the two SF devices still show appreciable emissions (10-15%) from the phosphorescent sensitizers, with the emission peak at around 510 nm from $Ir(ppy)_3$ and that at 583 nm from PQIr. This suggests an incomplete energy transfer of the triplet excitons from the phosphorescent sensitizers to the NIR fluorescent emitters, similar to that has been observed previously [10].

The *J-V* characteristics for NIR OLEDs based on BEDOT-$TQMe_2$ and BEDOT-BBT are shown in Figure 3(a). While both NF devices based on two DAD oligomers exhibit turn-on voltages of approximately 2.7 V, the two SF devices have slightly higher turn-on voltage of 3.4 V, probably due to the charge trapping behavior of the phosphorescent sensitizers.

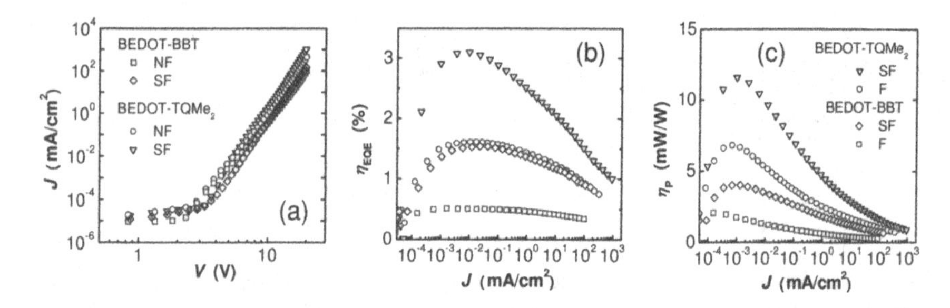

Figure 3. (a) Curret density, *J*, as functions of the voltage, *V*, for (normal) fluorescent (NF) and sensitized fluorescent (SF) OLEDs based on BEDOT-$TQMe_2$ and BEDOT-BBT. (b) External quantum efficiency, η_{EQE}, as functions of *J*, for the NF and SF OLEDs. (c) Power quantum efficiency, η_P, as functions of *J*, for the NF and SF OLEDs.

The EQE of these devices are shown in Figure 3(b). The BEDOT-$TQMe_2$ based NF device has a maximum EQE of η_{EQE} = (1.6 ± 0.2)% while the EQE for the BEDOT-BBT based NF device is lower, with a maximum of η_{EQE} = (0.51 ± 0.05)%. The maximum EQE of these two NF devices are approximately proportional to the corresponding fluorescent quantum yields of the DAD oligomers. The power efficiency is also higher for the BEDOT-$TQMe_2$ based device than the BEDOT-BBT based device, with a maximum of η_P = (7.0 ± 0.7) mW/W for the former device and (2.1 ± 0.2) mW/W for the latter.

When electrons and holes recombine in the EML to form excitons, both singlet and triplet excitons could be formed, and according to the spin statistics, only ¼ of the excitons are singlets whereas the remaining ¾ are triplets. In fluorescent OLEDs, only the singlet excitons are harvested on the emissive molecules as the radiative recombination process from the triplet excited state to the ground state is forbidden due to requirement of spin conservation. It has been shown that in fluorescent OLEDs, the incorporation of an appropriate phosphorescent dye molecule can induce mixing of singlet and triplet excitons, thereby funnel the triplet excitons formed on the host molecules to the fluorescent emitter and contribute to light emission [10]. Using such sensitized fluorescence device structure, the EQE of the BEDOT-TQMe$_2$ based SF device reaches a maximum of η_{EQE} = (3.1 ± 0.3)%, as shown in Figure 3(b). For the BEDOT-BBT based SF device, the maximum EQE is η_{EQE} = (1.5 ± 0.2)%. The maximum power efficiency is η_P = (12 ± 2) mW/W for the BEDOT-TQMe$_2$ based SF device, and is (4.0 ± 0.4) mW/W for the BEDOT-BBT based SF device, both approximately twice of that achieved in the NF devices. The maximum radiant emittances for BEDOT-TQMe$_2$ and BEDOT-BBT based SF devices, achieved at V = 20 V, are R = 19 and 2 mW/cm^2, respectively, which are more than three times higher than those of the corresponding NF devices.

In devices based on Pt-TPTNP, the spin-orbit coupling induced by the heavy Pt atom in the molecule leads to phosphorescent emission of the triplet excitons. This, in principle, will lead to a four-fold increase in quantum efficiency compared to the normal fluorescent devices. The EL spectrum of the OLEDs based on Pt-TPTNP is shown in Figure 4(a). The peak emission wavelength is at λ = 892 nm, and with a FWHM of approximately 50 nm, the emission peak is much narrower than the fluorescent devices based on the two DAD oligomers. The device shows a low turn-on voltage of approximately 2.2 V even though the BPhen ETL is rather thick (100 nm). Maximum radiant emittances of $R \approx 1.8$ mW/cm^2 are obtained at V = 17 V for the Pt-TPTNP based devices.

Figure 4(b) shows the current density dependencies of the EQE and power efficiencies of this phosphorescent device. The EQE is relatively constant at low current densities and reaches a maximum of η_{EQE} = (3.8 ± 0.3)% at $J \approx 0.1$ mA/cm^2; however, at $J > 1$ mA/cm^2, it decreases significantly with the increase of the current density, probably due to the triplet-triplet exciton annihilation process. The maximum power efficiency of the device is η_P = (19 ± 3) mW/W, achieved at low current densities ($J \approx 10^{-3}$ mA/cm^2). Compared with the NIR OLEDs based on BEDOT-BBT which show peak emission at 815 nm, the Pt-TPTNP based phosphorescent devices not only show longer wavelength emission, but exhibit maximum power efficiencies approximately 3-10 times higher than that of the NF OLEDs, or approximately 2-5 times higher than that of the SF devices.

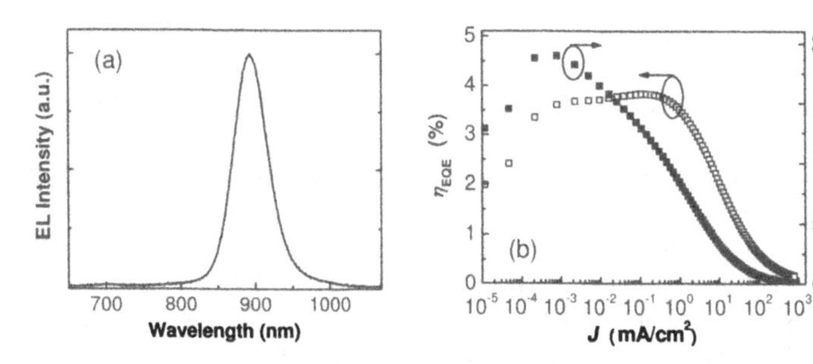

Figure 4. (a) Normalized EL spectra for OLEDs based on Pt-TPTNP. (b) External quantum efficiency, η_{EQE}, and power efficiency, η_P, as functions of J, for OLEDs based on Pt-TPTNP.

CONCLUSIONS

Here we demonstrate that low-gap DAD oligomers are good candidate materials for use in OLEDs to achieve tunable and efficient NIR emission. A maximum external quantum efficiency of 1.6% and a maximum power efficiency of 7.0 mW/W were achieved in devices based on BEDOT-TQMe$_2$, whose emission peaks at 692 nm. With a stronger acceptor in the molecule, longer wavelength NIR emissions peaked at 815 nm were achieved in BEDOT-BBT based devices, although the efficiencies were approximately three times lower due to the significantly lower fluorescent quantum yield of emitter molecule. Using the sensitized fluorescence device structure, the efficiencies were increased by two to three times, and we achieved the maximum efficiencies of $\eta_{EQE} = 3.1\%$ for BEDOT-TQMe$_2$ based devices, and $\eta_{EQE} = 1.5\%$ for BEDOT-BBT based devices.

We further demonstrate the use of a phosphorescent emitter, Pt-TPTNP, in OLEDs to further shift the emission wavelength into the NIR spectrum region and enhance the efficiencies. The peak emission wavelength at 892 nm, and maximum efficiencies of $\eta_{EQE} =$ 3.8% and $\eta_P = 19$ mW/W are finally achieved by the Pt-TPTNP based NIR OLEDs.

The authors gratefully acknowledge financial support from the U.S. Defense Advanced Research Projects Agency (DARPA) and the U.S. Army Aviation and Missile Research, Development, and Engineering Center (AMRDEC) (Project No. W31P4Q-08-1-0003) and the Office of Naval Research (Project No. N00014-05-1-0373) for this work.

REFERENCES

1. L. H. Slooff, A. Polman, F. Cacialli, R. H. Friend, G. A. Hebbink, F. C. J. M. van Veggel, and D. N. Reinhoudt, Appl. Phys. Lett. **78**, 2122 (2001).
2. N. Tessler, V. Medvedev, M. Kazes, S. Kan, and U. Banin, Science **295**, 1506 (2002).
3. S. A. Priola, A. Raines, and W. S. Caughey, Science **287**, 1503 (2000).
4. R. G. Sun, Y. Z. Wang, Q. B. Zheng, H. J. Zhang, and A. J. Epstein, J. Appl. Phys. **87**, 7589 (2000).
5. B. S. Harrison, T. J. Foley, M. Bouguettaya, J. M. Boncella, J. R. Reynolds, K. S. Schanze, J. Shim, P. H. Holloway, G. Padmanaban, and S. Ramakrishnan, Appl. Phys. Lett. **79**, 3770 (2001).
6. Y. Sun, C. Borek, K. Hanson, P. I. Djurovich, M. E. Thompson, J. Brooks, J. J. Brown, and S. R. Forrest, Appl. Phys. Lett. **90**, 213503 (2007).
7. T. A. Skotheim and J. R. Reynolds, *Handbook of Conducting Polymers*, 3rd ed. (CRC, New York, 2007).
8. T. T. Steckler, K. A. Abboud, M. Craps, A. G. Rinzler, and J. R. Reynolds, Chem. Commun. (Cambridge) **2007**, 4904.
9. B. C. Thompson, L. G. Madrigal, M. R. Pinto, T.-S. Kang, K. S. Schanze, and J. R. Reynolds, J. Polym. Sci., Part A: Polym. Chem. **43**, 1417 (2005).
10. M. A. Baldo, M. E. Thompson, and S. R. Forrest, Nature **403**, 750 (2000).
11. Y. Yang, R. T. Farley, T. T. Steckler, S.-H. Eom, J. R. Reynolds, K. S. Schanze, and J. Xue, Appl. Phys. Lett. **93**, 163305 (2008).
12. S. R. Forrest, D. D. C. Bradley, and M. E. Thompson, Adv. Mater. **15**, 1043 (2003).

Mater. Res. Soc. Symp. Proc. Vol. 1154 © 2009 Materials Research Society 1154-B05-118

Three-Dimensional Anisotropic Electronic Properties of Solution Grown Organic Single Crystals Measured by Space-Charge Limited Current (SCLC)

Beatrice Fraboni[1] , Alessandro Fraleoni-Morgera[2] and Anna Cavallini[1]
[1] Dipartimento di Fisica, Università di Bologna, viale Berti Pichat 6/2, 40127 Bologna, Italy
[2] Sincrotrone Trieste, Strada Statale Km 163.5, 34102 Basovizza (Trieste), Italy

ABSTRACT

Organic single crystals offer the interesting and unique opportunity to investigate the intrinsic electrical behaviour of organic materials, excluding hopping phenomena due to grain boundaries and structural imperfections. Their structural asymmetry permits also to investigate the correlation between their three-dimensional order and their charge transport characteristics. Here we report on millimeter-sized, solution-grown organic single crystals, based on 4-hydroxycyanobenzene (4HCB), which exhibit three-dimensional anisotropic electrical properties along the three crystallographic axes *a*, *b* (constituting the main crystal flat face) and *c* (the crystal thickness), measured over several different samples. The carrier mobility was determined by means of space charge limited current (SCLC) and air-gap field effect transistors fabricated with 4HCB single crystals and the measured values well correlate with the structural packing anisotropy of the molecular crystal. A differential analysis of SCLC curves allowed to determine the distribution and the concentration of the dominant electrically active density of states within the gap.

INTRODUCTION

Organic semiconductors are considered promising materials for implementing low-cost and large–scale produced electronics [1,2] and are receiving a large attention because of their potential applications, spanning from OLEDs to plastic photovoltaics and to bio-chemical sensors. However, the electronic transport properties of these materials are still not fully understood, and organic single crystals (OSCs) may represent model materials for assessing the charge transport mechanisms, thanks to their high purity and molecular order. Moreover, their structural asymmetry permits to investigate the correlation between their three-dimensional molecular stacking order and their charge transport characteristics [3,4].Recent studies showed that macroscopic crystals of rubrene present a two-dimensional electrical anisotropy [5,6] and single crystal organic transistors exhibited up to now the best performances in terms of charge carriers mobility, reaching time-of-flight-measured values as high as 400 $cm^2V^{-1}s^{-1}$, and FET-measured mobilities up to several $cm^2V^{-1}s^{-1}$.[3,5,7] In this view, the prospect of developing macroscopic (millimeter-sized), self-standing crystals, suitable for being manipulated and selectively deposed on any surface and in any position with respect to existing electrodes, is very attractive.

The more widely diffused method for obtaining crystals suited for these investigations is vacuum-based deposition, even if macroscopic organic crystals may be easily grown also from solution, permitting a considerable degree of control over the final crystal characteristics, in terms of dimensions and even of the developed crystallographic phase. These techniques may be applied also to semiconducting organic crystals suitable for applications in FETs, as demonstrated by recent reports [8].

In this work we report on millimeter-sized, solution-grown organic single crystals, based on 4-hydroxycyanobenzene (4HCB), which exhibited three-dimensional anisotropic electrical properties along the three crystallographic axes *a, b* (constituting the main crystal flat face) and *c* (the crystal thickness), measured over several different samples (Figure 1). Air-gap FET devices were used to estimate the carrier mobility along the two main axes *a* and *b*, reaching top values up to $8x10^{-2}$ and $9x10^{-3}$ $cm^2V^{-1}s^{-1}$ for μ_a and μ_b, respectively, while along the crystal thickness, axis *c*, the mobility was determined by means of space charge limited current (SCLC) measurements, which delivered a maximum value μ_c of $2x10^{-4}$ $cm^2V^{-1}s^{-1}$.

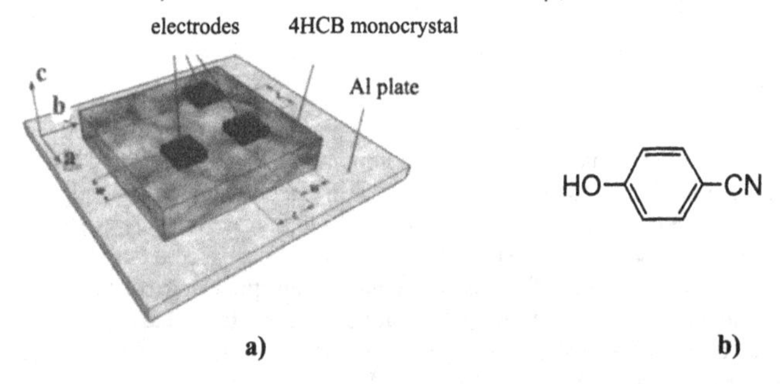

Figure 1: a) Schematic layout of the device geometry used for SCLC measurements, **b)** molecular structure of 4-hydroxycyanobenzene (4HCB).

EXPERIMENT

Solution-grown 4HCB crystals were obtained with a procedure based on a progressive and controlled evaporation of the solvent (ethylic ether), allowing to tune the dimensions of the 4HCB crystals from one to a few mm of face side and from1 mm down to about 150 μm of thickness. FTIR spectroscopy analyses were carried out before electrical characterization and no detectable traces of impurities were found. UV-Vis measurements delivered an optical bandgap of 296 nm (corresponding to 4.19 eV). To verify the crystalline structure of the solution-grown 4HCB crystals, X-ray diffraction analyses on some of the as-prepared crystals were conducted. The already known structure and morphological phase were confirmed.

All the electrical measurements reported in this paper were conducted in the dark, in air and at room temperature with a Keithley Source-Meters 2400 and Electrometers 617 and 6512. No significant hysteresis or time relaxation were observed. The ohmic contacts on 4HCB were fabricated using silver epoxy (Epo-Tek E415G) at a fixed distance, in order to form pairs of contact pads orthogonal to each other (Figure 1). As we already reported [9], the performance of the air-gap FETs fabricated with these contacts assessed the p-type conductivity of the material and the good hole injecting properties of the electrical contacts.

DISCUSSION

To investigate the transport anisotropy of 4HCB crystals we have carried out current-voltage analyses along the planar axes *a* and *b* using air-gap FETs, that we fabricated using 4HCB crystals as the active channel. Crystals, already contacted as described above, were placed directly on an Al gate electrode (Fig. 1a), leaving as a dielectric only a thin air layer [6,10]. It has been suggested that air-gap OFETs may provide a mobility measurement tool able to minimize interface effects induced by the gate dielectric [6,11], and may allow the measurement of almost bulk-like mobility.

To complete the three-dimensional characterization of the transport properties of 4HCB crystals, we have carried out current-voltage analyses along axis *c*, using the Space Charge-Limited Current (SCLC) method [12], a relatively easy-to-handle room-temperature investigation tool that allows to measure bulk-like mobility. SCLC curves have a symmetric character for positive and negative biases and are characterized by a linear region (ohmic) followed by a Space Charge Limited (SCL) one and finally by the Trap-Filled-Limited (TFL) region, when a steep increase of the current occurs. A typical SCLC curve measured along axis *c* is reported in Figure 2.

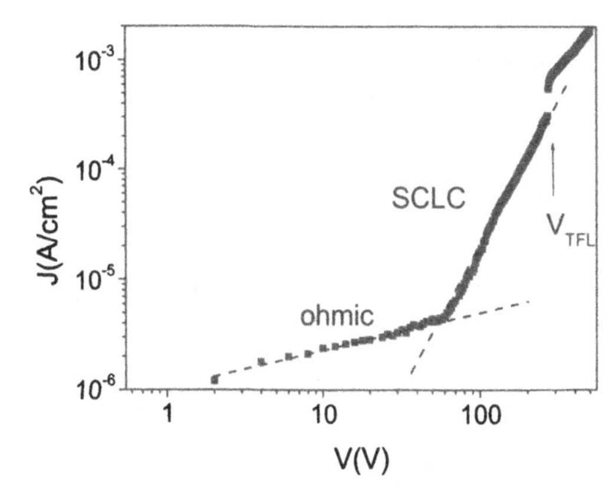

Figure 2: Typical Space Charge Limited Current curve measured in a 4HCB crystal along axis c

The vertical transport geometry has been used for SCLC measurements, i.e. the sandwich-like one (with electrical contacts on opposite sides of the crystal) (see Figure 1) [12], where the transport is assumed to be one-dimensional and follows the Mott and Gurney model:

$$J = \frac{9}{8}\varepsilon\mu\frac{V^2}{L^3} \qquad (1)$$

where J is the current density for the applied voltage V, L is the thickness of the organic layer (electrode separation), ε is the dielectric constant of the material and μ is the carrier mobility. The intrinsic-like carrier mobility μ_{SCLC} can be obtained from equation (1) calculated at $V=V_{TFL}$.

We have measured FET mobilities along axes *a* and *b* and SCLC curves along axis *c* over many different (in size and thickness) crystals, and the results are reported in Figure 3. A clear and reproducible anisotropy can be observed in the mobility values measured along the three crystal axes for several different crystals. The average mobilities measured along the three axes are 3×10^{-2} cm^{2}V^{-1}s^{-1} along *a*, 6×10^{-3} cm^{2}V^{-1}s^{-1} along *b* and 4×10^{-6} cm^{2}V^{-1}s^{-1} along *c*.

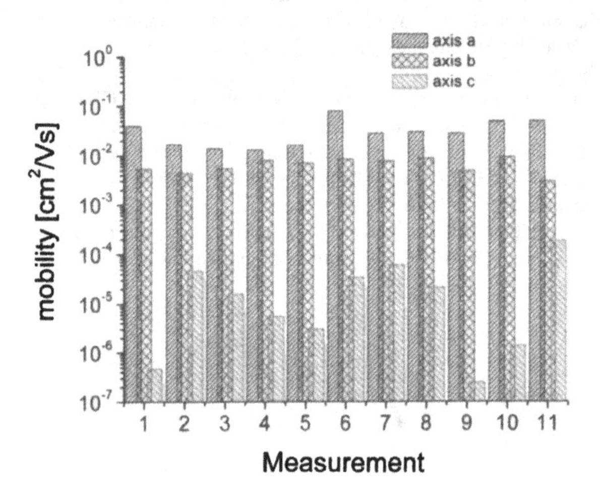

Figure 3: Charge carrier mobilities measured with the air-gap FET (axes a and b) and with the SCLC method (axis c) along the three spatial directions of many 4HCB crystals.

The measured electrical anisotropy well correlates with the crystal packing structure of 4HCB. In fact, the 4HCB lattice structure checked by means of X-ray diffraction measurements assessed that the packing along the *b* axis provides a moderate $\pi-\pi$ stacking between the benzenic rings, with a ring-to-ring distance of 10.738(2) Å, while along the axis a the inter-ring $\pi-\pi$ stacking distance is 9.202(2) Å, about 1.5 Å smaller than in the case of axis *b*, in good agreement with the higher mobility found along axis a. Finally, the structure along the *c* axis grows in a complex pattern of parallel spirals, with no appreciable overlapping of benzenic rings, again in agreement with the much lower mobilities found along axis *c* [8,9].

Another interesting information that can be extracted from SCLC analyses is the concentration and the energy distribution of the dominant electrically active traps in the band gap. The shape of the SCLC curve reflects the increment of the space charge with respect to the shift of the Fermi level as a function of the applied bias, and thus mirrors the energy dependence of the density of states (DOS) distribution. By analyzing the SCLC curves with a differential analysis method developed by Nespurek and Sworakowky [13], it is possible to extract an energy distribution of states $h(E_F)$ directly from a single current-voltage curve [14-16]. We have hence analyzed the SCLC curves obtained along the crystal axis *c* according to this model, and we have obtained the DOS distributions shown in figure 4. The DOS distribution is centered at an energy of 0.44 eV, with an average concentration of deep states of approx. 10^{13}cm^{-3}. Trap states at comparable energies have been recently reported along one of the axes of rubrene single crystals and in pentacene [14,16,17]. Further evidence on the presence of deep electrically active states in 4HCB crystals is provided by spectral photocurrent analyses, previously reported on 4HCB crystals [9], that indicate the occurrence of a strong electrically active band of defects located at approx. 0.50 eV from the HOMO–LUMO gap, a value that is in good agreement with the results obtained from the above reported SCLC analyses on the DOS energy distribution. Since 4HCB crystal are p-type conductors [8,9] and SCLC allows to study the behaviour of majority carriers, the observed deep band of states is attributed to hole traps.

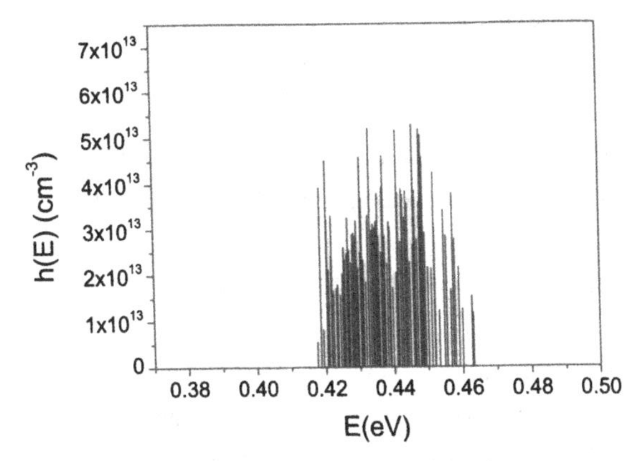

Figure 4: Density of states distribution obtained from SCLC analyses along axis *c* of 4HCB crystal.

CONCLUSIONS

In summary, millimeter-sized, dimension-tunable, self-standing single crystals of 4-hydroxy-cyanobenzene (4HCB) grown from solution exhibited a charge transport anisotropy related to all the crystallographic axes, namely *a*, *b* and *c*.

In particular, by air-gap FET devices (which allowed to classify the 4HCB crystal as a *p*-type material in the used experimental conditions), we measured mobilities μ_a and μ_b, along axes *a* and *b* of approximately $3x10^{-2}$ and $6x10^{-3}$ $cm^2V^{-1}s^{-1}$, respectively (values averages over several samples). Space charge limited current (SCLC) experiments conducted along the *c* axis, determined a mobility μ_c in the range of $4x10^{-6}$ $cm^2V^{-1}s^{-1}$ (average over several samples). The observed 3D anisotropy of the bulk-like mobility is well correlated with the structural anisotropic packing of the molecular crystal. Space charge limited current (SCLC) analyses along the *c* axis, revealed a density of states distribution of hole traps located at 0.4-0.5 eV from the HOMO-LUMO edge, with a concentration of about 10^{13} cm^{-3}. The found three-dimensional electrical anisotropy indicates that the solution growth technique can be regarded as a viable tool to investigate the electrical properties of organic single crystals along all the lattice axes.

ACKNOWLEDGMENTS

The authors acknowledge financial support by the Italian Research Ministry, under the project PRIN 2006

REFERENCES

1. S. R. Forrest, *Nature* **428**, 911 (2004)
2. A. Facchetti, *Mater. Today* **10**, 28 (2007)
3. R.de Boer, M.Gershenson, A.Morpurgo, V.Podzorov, *Phys.Status Solidi* **A 201**, 1302 (2004)
4. G. Horowitz, F. Garnier, A. Yassar, R. Hajlaoui, F. Kouki *Adv. Mater.,* **8**, 52 (1996)
5. R. Zeis, C. Besnard, T. Siegrist, C. Schlockermann, X. Chi, C. Kloc, *Chem. Mater.*, **18**, 244 (2006)
6. E. Menard, V. Podzorov, S.-H. Hur, A. Gaur, M. E. Gershenson, J. A. Rogers *Adv.* Mater. ,**16**, 2097 (2004)
7. O. D. Jurchescu, J. Baas, T. T. M. Palstra Appl. Phys. Lett. **84**, 3061 (2004)
8. a) S. C. B. Mannsfeld, J. Locklin, C. Reese, M. E. Roberts, A. J. Lovinger, Z. Bao, Adv. Funct. Mater. **17**, 1617 (2007); b) B.Fraboni, I.Mencarelli, L Setti, C.Femoni, R.DiPietro, A.Cavallini, A.Fraleoni, Adv.Mater. 2009 in press
9. B. Fraboni, R. DiPietro, A. Castaldini, A. Cavallini, A. Fraleoni Morgera, L. Setti, I. Mencarelli, C. Femoni, Org. El., **9**, 974 (2008)
10. A. F. Stassen, R. W. I. De Boer, N. N. Losad, A. F. Morpurgo, Appl. Phys. Lett. **85,** 3899 (2004)
11. J.Takeya et al., Phys. Rev. Lett. **98**, 196804 (2007)
12. M.Lampert and P. Mark, in: Current Injection in Solids, Academic Press, New York, 1970.
13. J.Sworakowski, S.Nespurek, *Vacuum* **38**, 7 (1989)
14. D.Braga, N.Battaglini, A.Yassar, G.Horowitz, *Phys. Rev. B* , **77**, 115205 (2008)
15. T.Cesca, A. Gasparotto, B. Fraboni *Appl. Phys. Lett.* **93**, 102114 (2008)
16. C.Krellner et al., *Phys. Rev. B,* **75**, 245115 (2007)
17. D.Lang et al., *Phys. Rev. Lett.* **93,** 76601 (2004)

Understanding Interfaces

Mater. Res. Soc. Symp. Proc. Vol. 1154 © 2009 Materials Research Society 1154-B06-07

Formation and Electrical Interfacing of Nanocrystal-Molecule Nanostructures

Claire Barrett[1], Gaëtan Lévêque[2], Hugh Doyle[1], Donocadh P. Lydon[3], Gareth Redmond[1§], Trevor R. Spalding[3] and Aidan J. Quinn[1*]
[1]Nanotechnology Group, Tyndall National Institute, Lee Maltings, Prospect Row, Cork, Ireland
[2]Photonics Theory Group, Tyndall National Institute, Lee Maltings, Prospect Row, Cork, Ireland
[3]Department of Chemistry, University College Cork, Cork, Ireland
[§]Current Address: School of Physics, University College Dublin, Belfield, Dublin 4, Ireland
*Corresponding Author. Email: aidan.quinn@tyndall.ie

ABSTRACT

The formation of nanocrystal-molecule-nanocrystal nanostructures via controlled mixing of Au nanocrystals and bifunctional Re linkers is reported. UV-visible extinction data, coupled with histogram analysis of nanostructures measured using scanning electron microscopy has shown a characteristic optical response at wavelengths close to 600 nm following formation of dimer and trimer nanostructures. Directed assembly processes based on dielectrophoretic trapping have also been developed for electrical interfacing of these nanostructures between top-down nanoelectrode pairs for electrical characterization.

INTRODUCTION

Recent developments in the design and synthesis of nanoscale building blocks as active elements for information storage, biosensing and photoconductive devices have the potential to revolutionize several emerging technology markets across multiple sectors including healthcare, printable electronics, security and energy conversion. Functional organic molecules (~1 – 2 nm length scale) are attractive candidates as building blocks due to their composition-, size- and structure-dependent electronic properties, the ability to design and manipulate these properties *via* low-cost chemical synthesis, and finally the potential for formation of ordered structures through (bio)-molecular recognition and self-assembly [1, 2]. Ligand-stabilized inorganic nanocrystals (~2 – 50 nm core diameter) also represent attractive candidates, both as building blocks with novel functionality (e.g., single charge tunneling or quantum confinement) and also as terminals or nodes to bridge the gap between length scales accessible via top-down lithography (~ 25 – 50 nm) and molecular length scales [3-6]. The plasmonic properties of nanocrystal-molecule nanostructures have also attracted considerable attention [7-10]. We present recent results on solution-based formation of nanocrystal-molecule-nanocrystal nanostructures and investigation of the novel plasmonic properties of these nanostructures via measurement and simulation. Further, we report on directed assembly of nanostructures at contact nanoelectrodes and initial investigations of charge transport in these "few-molecule" devices.

EXPERIMENT

Citrate-stabilized Au nanocrystals with core diameters $d = 20 \pm 2$ nm were purchased from British Biocell Ltd. Ultrapure deionized water with resistivity 18.2 MΩ cm (ELGA PURELAB Ultra) was used in all experiments. To improve the stability of the nanocrystals, 3 mM aqueous tri-sodium citrate was added prior to use. The bi-functional linker molecule, Re_2(2',6'-dimethylacetanilido)$_4$(NCS)$_2$ was synthesized as previously described and dissolved in acetone (ACS grade); see Figure 1a for a schematic [11]. These Re_2(DMAA)$_4$(NCS)$_2$ linkers were selected as candidates for this study due to their miscibility (in acetone) with aqueous dispersions of gold nanocrystals. Routes for formation of nanostructures using conjugated "rigid rod" oligo (phenylene ethynylene) derivatives as linkers and citrate-stabilized nanocrystals phase-transferred to organic solvents are currently being developed.

For the experiments reported here, a 3 mL aliquot of diluted $d = 20$ nm Au nanocrystal solution (3.5×10^{11} nanocrystals/mL) was decanted into a clean cuvette with a single magnetic stir-bar and the UV-Visible extinction spectrum was recorded using an Agilent 8453 spectrophotometer. For each cycle of linker addition, the cuvette was placed on a stirrer plate and set to stir at a moderate rate. Aliquots (5 – 10 μL each) of 10 μM linker solution were added to the nanocrystal solution at 1 – 3 minute time intervals. After mixing, the cuvette was re-inserted into the spectrometer and the UV-Visible extinction was recorded at regular intervals (1 – 10 minutes).

Scanning electron microscopy protocols (JSM-6700F, JEOL UK Ltd.) were developed to record and statistically analyze the populations of monomers, dimers, trimers and higher order *n*-mers formed. Samples were prepared by depositing a dilute drop of the *n*-mer solution onto a silicon chip and wicking away the solution. Control samples were also prepared using unmodified Au nanocrystals from the same batch used to prepare the *n*-mers. For each substrate, approximately 100 nanostructures were counted at 10 – 20 random locations.

A wafer-level nanoelectrode fabrication process was developed for two-terminal electrical interfacing of nanostructures. A range of electrode geometries and gap sizes with critical dimensions down to 25 nm were fabricated using electron-beam lithography (JBX-6000FS, JEOL UK Ltd.), metal evaporation (3 nm Ti, 15 nm Au) and lift-off. Micron scale tracks and larger contact pads were fabricated using optical lithography, metal evaporation (10 nm Ti, 200 nm Au) and lift-off. Following fabrication, the dice were solvent cleaned in acetone and isopropanol, ashed in O_2 plasma (March Plasmod, 0.5 Torr, 20 W, 3 minutes) and stored under ethanol until used.

Dielectrophoretic trapping was employed for directed assembly of unmodified Au nanocrystals or nanocrystal-molecule nanostructures between nanoelectrodes. A drop (6 μL) of nanocrystal solution was placed on the central region of the chip. An a.c. voltage with frequency (f) in the range 1 MHz $\leq f \leq$ 10 MHz and peak-to-peak amplitude ($V_{p\text{-}p}$) in the range $0.1 \leq V_{p\text{-}p} \leq 1.8$ V was applied for 60 – 120 seconds to trap the nanostructures. For each set of values for the trapping parameters (f, $V_{p\text{-}p}$), the process was repeated across 10 sets of nanoelectrodes on the same chip. Room temperature current-voltage (I-V) characteristics for each device were measured in a vacuum probe-station (Desert Cryogenics TTP-4-HF) interfaced to a parameter analyzer (Agilent E5270B).

DISCUSSION

Figure 1a shows a schematic of the formation process, where mixing can lead to formation of structures containing one nanocrystal ("monomers", $n = 1$), two nanocrystals ("dimers", $n = 2$) and higher order *n*-mers. Figure 1b shows a selection of high-resolution scanning electron microscopy (SEM) images of *n*-mers drop-deposited onto silicon substrates following formation in solution through mixing of citrate-stabilized Au nanocrystals (with core diameters, $d = 20 \pm 2$ nm) and Re_2(DMAA)$_4$(NCS)$_2$ linkers. The separation between the sulphur end groups on each linker has been calculated from X-ray crystallography data to be 1.2 nm [11]. It is important to note that the description of the n-mer does not specify the number of molecules attached to each nanocrystal.

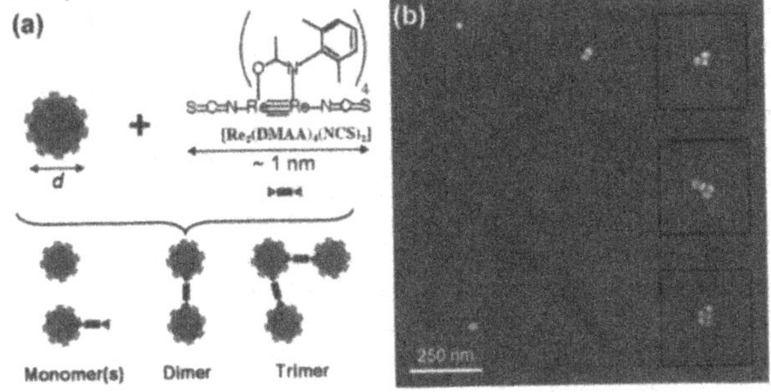

Figure 1 (a) Schematic (not to scale) of nanocrystal-molecule "n-mer" nanostructures formed by mixing Au nanocrystals (d = 20 nm) with bi-functional Re linker molecules. (b) SEM image showing 2 monomers and 1 dimer. Insets: Examples of a trimer, tetramer and pentamer, respectively.

Unmodified d = 20 nm Au nanocrystals showed the characteristic isolated nanocrystal extinction with peak intensity, λ_{max}, at 523 nm, corresponding to the well-know plasmon resonance in gold nanocrystals; see Figure 2a. Bifunctional Re linker was added in aliquots to the nanocrystals over a 2.5 hour period (with a final molecule:nanocrystal ratio ~ 800:1), during which time 30 spectra were recorded.

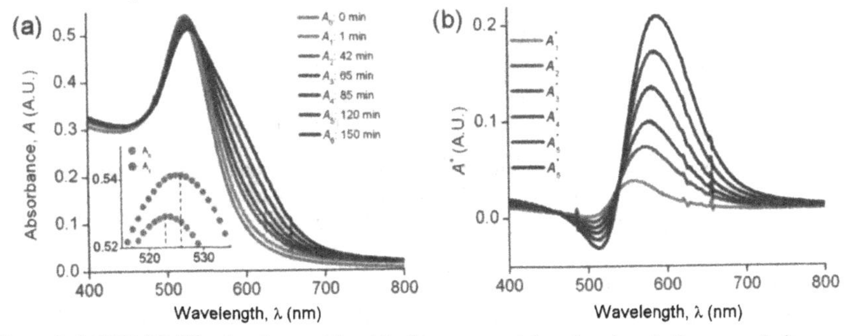

Figure 2. (a) UV-Visible absorbance (A_1- A_6) of a nanocrystal-molecule solution recorded versus wavelength (λ) over 2.5 hours. A_0 is the absorbance measured for the starting solution of citrate-stabilized Au nanocrystals. The inset shows the peak positions of A_0 and A_1, a 2-3 nm red-shift can be observed. (b) Relative absorbance (A^*) of the nanocrystal-molecule solutions shown in (a), following subtraction of the citrate spectrum according to: $A_i^*(\lambda) = A_i(\lambda) - A_0(\lambda)$, i = 1...6.

Figure 2a shows a subset of the recorded extinction spectra ($A_1, A_2 \ldots A_6$) chosen to show the evolution of the optical response. Almost immediately after addition of the linkers (A_1, 1 minute), a red-shift ($\Delta\lambda \approx$ 2-3 nm) of the peak close to 523 nm could be observed, likely due to initial adsorption of the Re linkers at each nanocrystal surface [12]. During the experiment the emergence of a second feature was observed close to 600 nm. The evolution of this shoulder can be highlighted by subtracting the absorbance data of unmodified Au nanocrystals, i.e., A_0 in Figure 2a, from the measured absorbance (A_i) at each time interval; see Figure 2b.

Figure 3 shows a histogram (dark blue data) of the *n*-mer nanostructure distribution for the solution whose extinction spectrum (A_6) is shown in Figure 2a. The data, totaling 480 nanostructures, were extracted from analysis of 67 SEM images acquired at different locations across 4 substrates. The error bars, which show the 95% confidence interval for the analysis (± 1.96 σ) confirm the reliability of the method to determine the distribution of nanostructures in a solution. The distribution comprises roughly 66% monomers, 20% dimers, 9% trimers and low incidences of higher order *n*-mers. Only 2 nanocrystal-molecule nanostructures were observed with $n > 7$. Control experiments on unmodified nanocrystal solutions prepared in the same manner but without addition of linker molecules yielded the pale pink histogram data shown in Figure 3 (112 nanostructures counted). Over 95% monomers were observed for the bare nanocrystals with a low incidence of dimers and negligible incidence of higher order *n*-mers. The statistics indicate that the nanostructures observed using SEM were formed in solution and not as a result of aggregation during drop-deposition onto the substrate or solvent evaporation.

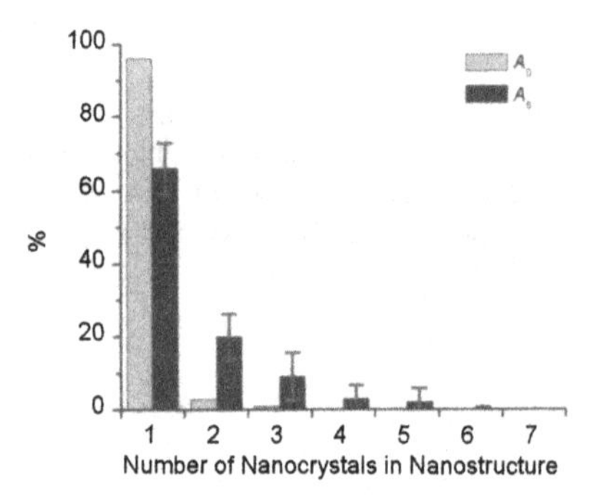

Figure 3. Histogram (dark blue) showing the distribution of *n*-mer nanostructures extracted from SEM data measured for the nanocrystal-molecule solution whose extinction data (A_6) is shown in Figure 2a. Control data (pale pink) for an unmodified Au nanocrystal solution (A_0 from Figure 2a)

The *n*-mer distribution extracted from SEM data suggests that the measured extinction feature close to 600 nm in Figure 2 arises from the optical response of nanocrystal-molecule dimer or trimer nanostructures. Simulations and optical scattering experiments on fabricated and synthesized dimer nanostructures have revealed the existence of a second peak, considerably red-shifted with respect to the plasmon peak for isolated spherical nanostructures [7-10]. Simulations are currently being developed using the Generalized Multiparticle Mie method to model the optical extinction of nanocrystal dimers for electric field polarizations parallel ($E_{//}$) and perpendicular ($E_{\perp}$) to the dimer axis, respectively [13]. Initial results for dimers comprising $d =$ 20 nm Au nanocrystals with edge-edge separation ~ 1.2 nm (expected length for Re linkers) show two main peaks: one close to 520 nm and a second longer-wavelength peak close to 600 nm; see Figure 4. This longer wavelength feature is more prominent for case where the field polarization is parallel to the dimer axis ($E_{//}$), suggesting that the shoulder close to 600 nm in the measured extinction data shown in Figure 2 arises from a longitudinal excitation in the dimer nanostructures [7].

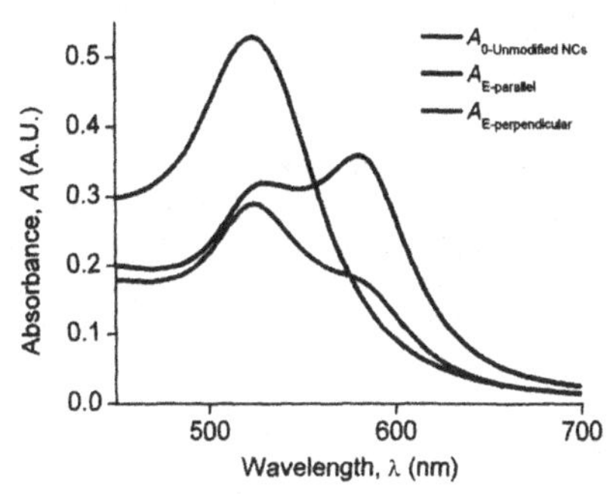

Figure 4. Modeled extinction spectra for both $E_{//}$ and $E_{\perp}$ field orientations for a dimer formed from $d = 20$ nm Au nanocrystals with edge-edge separation of 1.2 nm. The measured extinction spectrum of unmodified $d = 20$ nm Au nanocrystals (A_0 from Figure 2a) is shown for reference.

Nanoelectrode chips were prepared as outlined above and the nanocrystal-molecule nanostructures were trapped using dielectrophoresis ($V_{p\text{-}p} = 1.8$ V, $f_{DEP} = 1$ MHz, $t_{DEP} = 60$ s). Current-voltage (I_d -V_{ds}) measurements were acquired for all devices at room temperature in a vacuum probe station and devices were subsequently inspected using SEM to quantify the trapping process. Of the 40 electrode pairs used, 27 showed trapping events. Of these 27 devices, only 3 showed measurable current, i.e., $I_d > 2$ pA at $V_{ds} = 1$ V. These three devices yielded resistances of 2.5 ± 0.1 GΩ, 15.2 ± 0.9 GΩ and 147 ± 10 GΩ, respectively, calculated using linear regression of measured I_d -V_{ds} data in the range $0.2 \text{ V} < V_{ds} < 1$ V. Figure 5a shows a high resolution SEM image of the 2.5 GΩ device post *I-V*. A variety of *n*-mer nanostructures and other aggregates can be observed bridging the electrodes. Figure 5b shows the measured current-voltage (I_d -V_{ds}) data for this device. The data exhibit slight non-linearity at high bias, consistent with tunneling [14].

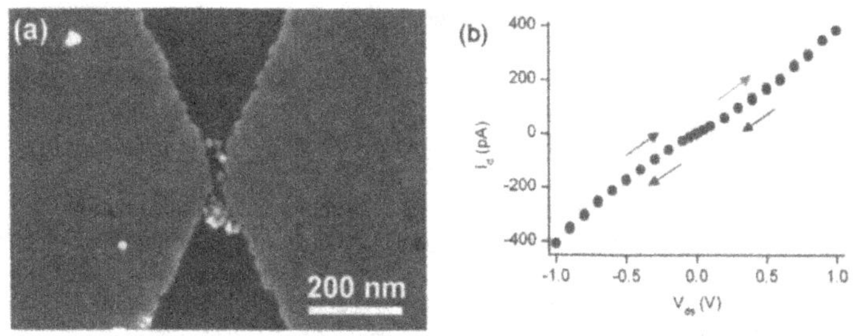

Figure 5 (a) SEM showing directed assembly of *n*-mer nanostructures at contact electrodes via dielectrophoresis. (b) I_d -V_{ds} data acquired for the device shown in (a) prior to SEM imaging, yielding a (high-bias) resistance, $R = 2.5 \pm 0.1$ GΩ.

Since the yield of devices with measurable current was quite low (3 out of 27), alternative contacting strategies are also being pursued, based on dielectrophoretic trapping of unmodified *d*

= 50 nm Au nanocrystals, I_d -V_{ds} characterization, immersion in the linker solution and re-measurement of I_d -V_{ds} data to monitor conductance changes. Initial results show a higher yield of measurable devices (~ 50%).

CONCLUSIONS

Protocols have been developed for formation of nanocrystal-molecule nanostructures. The novel plasmonic properties of these structures offer the potential for development of "in-line" optical techniques for monitoring nanostructure formation. Directed assembly protocols have also been developed for electrical interfacing of nanocrystal-molecule nanostructures. These results represent an important step towards development of scalable contacting methods for functional molecules.

ACKNOWLEDGMENTS

The authors acknowledge Yoav Gordin and Israel Bar-Joseph for technical discussions, and also Dan O'Connell, Brendan McCarthy and Richard Murphy (Tyndall Central Fabrication Facility) for technical assistance in the design and fabrication of the nanogap electrodes. This work was supported by Science Foundation Ireland (grant number RFP06-PHY005), the Irish HEA PRTLI Programmes (Cycles 3 & 4) and the European Union ("FUNMOL" project, 213382).

REFERENCES

1. J. R. Heath and M. A. Ratner, Phys. Today **56** (5), 43-49 (2003).
2. N. J. Tao, Nature Nanotechnology **1** (3), 173-181 (2006).
3. L. C. Brousseau, J. P. Novak, S. M. Marinakos and D. L. Feldheim, Adv. Mater. **11** (6), 447-449 (1999).
4. T. Dadosh, Y. Gordin, R. Krahne, I. Khivrich, D. Mahalu, V. Frydman, J. Sperling, A. Yacoby and I. Bar-Joseph, Nature **436** (7051), 677-680 (2005).
5. C. W. Chu, J. S. Na and G. N. Parsons, J. Am. Chem. Soc. **129** (8), 2287-2296 (2007).
6. J. Liao, L. Bernard, M. Langer, C. Schonenberger and M. Calame, Adv. Mater. **18** (18), 2444-2447 (2006).
7. T. Jensen, L. Kelly, A. Lazarides and G. C. Schatz, J. Clust. Sci. **10** (2), 295-317 (1999).
8. T. Atay, J. H. Song and A. V. Nurmikko, Nano Lett. **4** (9), 1627-1631 (2004).
9. C. Sonnichsen, B. M. Reinhard, J. Liphardt and A. P. Alivisatos, Nature Biotechnology **23** (6), 741-745 (2005).
10. P. Billaud, S. Marhaba, E. Cottancin, L. Arnaud, G. Bachelier, C. Bonnet, N. Del Fatti, J. Lerme, F. Vallee, J. L. Vialle, M. Broyer and M. Pellarin, J. Phys. Chem. C **112**, 978-982 (2008).
11. D. P. Lydon, T. R. Spalding and J. F. Gallagher, Polyhedron **22** (9), 1281-1287 (2003).
12. W. Haiss, N. T. K. Thanh, J. Aveyard and D. G. Fernig, Anal. Chem. **79** (11), 4215-4221 (2007).
13. Y.-l. Xu, Appl. Opt. **34** (21), 4573-4588 (1995).
14. J. G. Simmons, J. Appl. Phys. **34** (6), 1793-1803 (1963).

Charges and Transport

Mater. Res. Soc. Symp. Proc. Vol. 1154 © 2009 Materials Research Society 1154-B07-07

High Performance n-Type Organic Thin-Film Transistors With Inert Contact Metals

Sarah Schols[1,2], Lucas Van Willigenburg[1], Robert Müller[1], Dieter Bode[1,2], Maarten Debucquoy[1,2], Jan Genoe[1], Paul Heremans[1,2], Shaofeng Lu[3] and Antonio Facchetti[3]

[1] IMEC v.z.w., PME-LAE, Kapeldreef 75, 3000 Leuven, Belgium
[2] Katholieke Universiteit Leuven, ESAT-INSYS, Kasteelpark Arenberg 10, 3000 Leuven, Belgium
[3] Polyera Corporation, 8025 Lamon Avenue, Skokie, IL 60077, USA

ABSTRACT

Thin film growth by high vacuum evaporation of the n-type organic semiconductor 5, 5'''-diperfluorohexylcarbonyl-2,2':5',2'':5'',2'''-quaterthiophene (DFHCO-4T) on poly-(α-methylstyrene)-coated n^{++}-Si/SiO_2 substrates is investigated at various deposition fluxes and substrate temperatures. Film characterization by atomic force microscopy reveals typical Stransky-Krastanov growth. Transistors with Au source-drain top contacts and optimized DFHCO-4T deposition conditions attain an apparent saturation mobility of 4.6 cm^2/Vs, whereas this parameter is 100× lower for similar transistors with LiF/Al top contacts. We explain this lower performance by the formation of a thin interfacial layer with poor injection properties resulting from a redox reaction between Al and DFHCO-4T.

INTRODUCTION

Driven by potential applications of complementary logic, the field of electron-channel (n-type) organic thin-film transistors (n-OTFT) has recently gained a lot of attention. Using n-type materials such as fullerene[1], naphthalene[2], perylene[3] and oligothiophene[4] derivatives, high performance n-OTFTs have been reported. 5,5'''-diperfluorohexylcarbonyl-2,2':5',2'':5'',2'''-quaterthiophene (DFHCO-4T, Figure 1) is an example of such a promising electron-conducting organic semiconductor. Recently, DFHCO-4T was shown to obtain an electron field-effect mobility of 1.7 cm^2/Vs.[5] A top-contact geometry using Au source and drain contacts was used in these devices.

In this paper, we discuss the optimization of DFHCO-4T growth and compare the performance of DFHCO-4T transistors with different top-contact metals. While most n-OTFTs need low-workfunction metal contacts such as Mg, Ca or LiF/Al for efficient electron injection into the semiconductor lowest unoccupied molecular orbital (LUMO),[5,6] we find that DFHCO-4T n-OTFTs only function properly with contacts that have a low chemical reactivity such as Au and Ag. Electron mobilities as high as 4.6 cm^2/Vs are reported for Au top-contact transistors using optimized DFHCO-4T growth conditions. This fact is of high technological relevance because for use in complementary logic it is preferable to use a single type of source and drain metal for both the p-type and the n-type OTFTs. The reduced performance with easily oxidizable metals such as Al and Yb is explained as a consequence of an electron-transfer reaction occurring at the metal/DFHCO-4T interface.

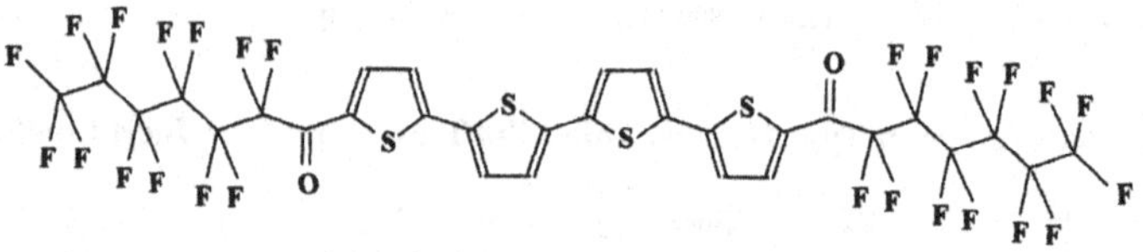

Figure 1. Molecular structure of DFHCO-4T (ActivInk™ 0800).

EXPERIMENTAL

Devices were fabricated on highly doped n^{++}-Si wafers with a 140-nm thick thermally grown SiO_2 layer (serving as the OTFT gate dielectric). Prior to the deposition of the organic semiconductor, the substrates were covered with a 5-nm thick poly-(α-methylstyrene) layer (PαMS, standard 700.000, Fluka) prepared by spin-coating a solution of 0.1 weight% PαMS in toluene at 4000 rpm and drying on a hotplate at 120°C for 1 min.[8] This procedure provides a high-quality, electron-trap free surface allowing excellent electron transport.[9] Thin DFHCO-4T films were deposited on these substrates by thermal vacuum evaporation ($p = 10^{-8}$ torr) of the material (ActivInk™ N0800, Polyera Corporation), which was used as received.

To obtain transistors, the structure was completed by evaporation of metallic top-contacts (100 nm Au or 0.8 nm LiF followed by 100 nm Al) onto the organic semiconductor through a shadow mask. The resulting device channel width was 2000 μm and the channel length varied between 50 and 200 μm. Immediately after fabrication the devices were transferred to a N_2 filled glove box (without exposure to air) for electrical characterization using an Agilent 4156C parameter analyzer. Thin film mobilities were calculated from saturation,[3] using

$$I_D = \frac{1}{2}\frac{W}{L}C_{ox}\mu(V_{gs} - V_T)^2 \tag{1}$$

where W is the width of the transistor, L is the transistor channel length, C_{ox} is the capacitance per unit area of the SiO_2 gate dielectric, μ is the mobility, V_{gs} is the gate-source voltage and V_T is the threshold voltage. The total capacity of the gate dielectric per area C_{ox} was calculated based on the relative dielectric constant ε_r and the thickness t of PαMS and SiO_2 using the following formula:

$$\frac{1}{C_{ox}} = \frac{1}{C_{SiO_2}} + \frac{1}{C_{P\alpha MS}} = \frac{t_{SiO_2}}{\varepsilon_0 \varepsilon_{r_{SiO2}}} + \frac{t_{P\alpha MS}}{\varepsilon_0 \varepsilon_{r_{P\alpha MS}}} \tag{2}$$

Capacitance-voltage measurements pointed out that the relative dielectric constant of thermally grown SiO_2 is 3.9, whereas ε_r=2.5 for PαMS, giving rise to a total capacity of the dielectric layer per area of $2.33\ 10^{-4}$ F/m^2.

RESULTS AND DISCUSSION

Atomic force microscopy (AFM) analysis of DFHCO-4T thin films reveals typical Stransky-Krastanov growth mode:[11] with the first few DFHCO-4T layers grown in a two-dimensional (2D) arrangement followed by a three-dimensional (3D) lattice. In the case of DFHCO-4T films, thick elongated needles (up to 100 nm long) are formed. Figure 2 illustrates how the deposition conditions affect the morphology of the DFHCO-4T films. An AFM image of a 50-nm thick (as measured by a quartz crystal monitor) DHFCO-4T film grown on top of PαMS at a low deposition flux (0.2 Å/s) while the substrate was kept at room temperature is shown in Figure 2(a). Under these conditions, a very rough morphology dominated by high peaks, which is typical for 3D growth, is observed. By increasing the deposition flux as well as by increasing the substrate temperature, the transition from 2D to 3D growth can be delayed. At high flux and high substrate temperature the film morphology becomes much smoother, characterized by large 2D grains with only a few 3D features (deposition rate of 4.5 Å/s and substrate temperature of 70 °C, Figure 2(f)). Consequently, these latter growth conditions are preferred for transistor fabrication since a reduced number of 3D features improve the quality and the homogeneity of the interface between the metal contact and the organic semiconductor

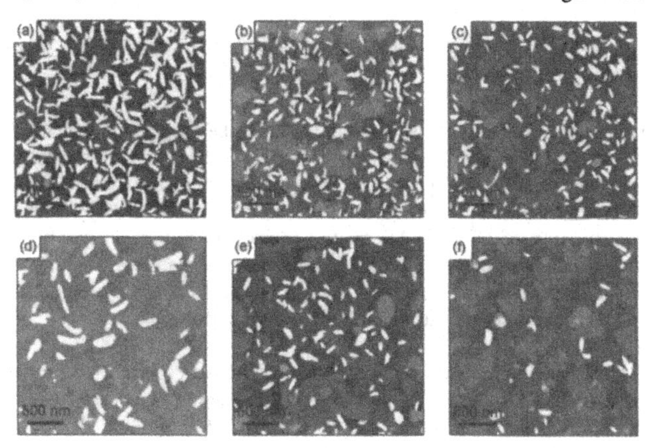

Figure 2. 2.5 μm x 2.5 μm AFM topography images of 50-nm thick DFHCO-4T films grown on PαMS at different deposition rates and different substrate temperatures: (a) flux = 0.2 Å/s and T_{sub} = 20 °C, (b) flux = 3 Å/s and T_{sub} = 20 °C, (c) flux = 4.5 Å/s and T_{sub} = 20 °C, (d) flux = 0.2 Å/s and T_{sub} = 70 °C, (e) flux = 3 Å/s and T_{sub} = 70 °C, and (f) flux = 4.5 Å/s and T_{sub} = 70 °C.

Figure 3(a) shows the output characteristics of a transistor with 130 μm channel length and Au top contacts. DFHCO-4T was deposited using the optimized growth conditions of Figure 2(f). An apparent field-effect electron mobility of 4.6 cm^2/Vs was measured. This field-effect mobility is higher compared to previously reported results[5] and we attribute this to the different growth conditions of the organic semiconductor layer DFHCO-4T. 20 transistors on 6 different samples, which were fabricated during different evaporation runs, were measured and the reproducibility of the results was quite good. The average mobility, calculated from transistors with channel lengths between 50 and 200 μm is 3.5 cm^2/Vs. The fact that electrons can be efficiently injected from Au into DFHCO-4T despite an injection barrier of ~1eV is probably related to the deep LUMO (3.96eV)[12] of DFHCO-4T and the fact that in a top-contact geometry there is a high gate-field that supports the injection of charges in a large source-gate overlap area[13], which makes it easier to overcome the electron injection barrier.

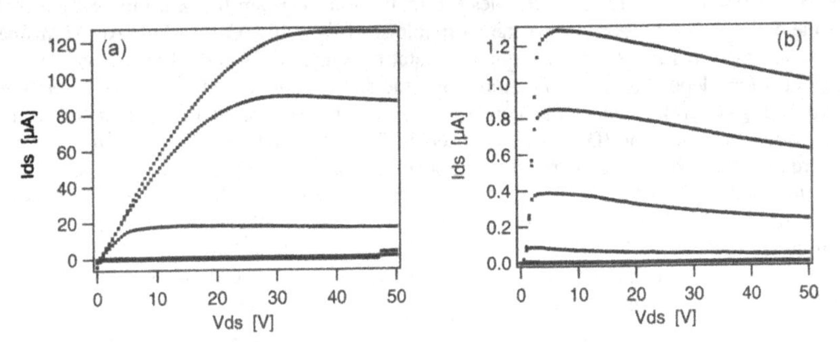

Figure 3. Output characteristics of DFHCO-4T transistors (W/L=2000/130) using (a) Au, and (b) LiF/Al top contacts.

In a next step, we checked whether also LiF/Al could be used as top contact material. The output characteristics of a DFHCO-4T transistor with 130 μm channel length and LiF/Al source-drain top contacts are presented in Figure 3(b). For this device, an apparent field-effect electron mobility of 0.03 cm^2/Vs was calculated. This value is more than two orders of magnitude lower than the electron mobility of DFHCO-4T Au top-contact transistors. In addition, the OTFTs show strongly non-ideal characteristics, as apparent from a superlinear dependence of the current on the drain voltage at low bias, the saturation of the output current at a more or less fixed drain-to-source bias, and a decrease of the current at higher drain voltages.

The low apparent field-effect mobility (i.e., low output currents) and the non-ideal characteristics of DFHCO-4T transistors with LiF/Al top contacts compared to Au top contacts point to the presence of a current-limiting contact problem in the former. A plausible cause of this is the occurrence of a redox reaction between the n-type semiconductor and the Al metal contact. In fact, it is well known that ketone groups – present in the DFHCO-4T chemical structure – are easily reduced into their radical anions (Figure 4(a)).[14] This electron transfer reaction is driven by the oxidation of a reactive metal, as for example Al (Figure 4(b))[15,16] and the reduction of DFHCO-4T. The overall redox reaction is depicted in Figure 4(c).

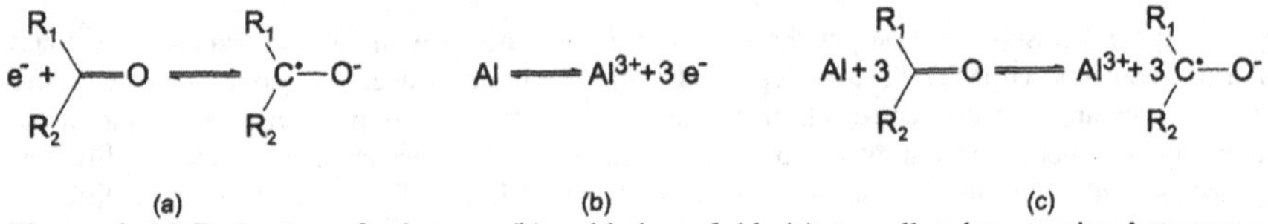

$$e^- + (R_1)(R_2)C{=}O \longrightarrow (R_1)(R_2)C^{\bullet}{-}O^- \quad \text{(a)}$$

$$Al \longrightarrow Al^{3+} + 3\,e^- \quad \text{(b)}$$

$$Al + 3\,(R_1)(R_2)C{=}O \longrightarrow Al^{3+} + 3\,(R_1)(R_2)C^{\bullet}{-}O^- \quad \text{(c)}$$

Figure 4. (a) Reduction of a ketone, (b) oxidation of Al, (c) overall redox reaction between a ketone and Al.

The standard free energy ($\Delta G°$, Gibbs free energy) of this overall redox reaction can be calculated by $\Delta G° = - n F E°$,[17] where n is the overall number of electrons exchanged between the oxidizing and reducing agents for the balanced redox equation, F the Faraday constant (96485 J/(V.mol), and E° the standard cell potential (the difference in standard electrode potentials (SEPs) of both electrochemical couples). Using a SEP of -1.90 V vs SCE for the Al/Al^{3+} couple[18,19] and -0.88 V vs SCE for DFHCO-4T,[12] gives $\Delta G°$= -296 kJ/mol. This indicates that the reaction between Al and DFHCO-4T is highly exergonic and thus thermodynamically spontaneous. The products of this reaction, an ionic salt and possibly dimerized ketone species, will form a thin interfacial layer between the unreacted DFHCO-4T and Al layers, hindering electron injection. In the case of Au this kind of reaction is impossible since the SEP's of the couples involving oxidation of Au are so high (+1.692 V vs NHE for Au/Au^{+} and +1.492 V vs NHE for Au/Au^{3+})[18] that a reaction with DFHCO-4T would be highly endergonic ($\Delta G°$=+225 kJ/mol in the case of Au/Au^{+} and +619 kJ/mol for Au/Au^{3+}). This explains the much better-behaved output characteristics for Au top-contact transistors compared to devices where LiF/Al is used as metal contact.

To find additional proof supporting the chemical reaction between Al and DFHCO-4T, transistors have been fabricated using DFHCO-4T as the organic semiconductor but with two other top-contact metals: Yb and Ag.[20] In the case of Yb, which is an easily oxidizable metal, similar output characteristics and low apparent field-effect mobility (0.06 cm^2/Vs) as Al/LiF transistors were measured.[20] The output characteristics of Ag top-contact transistors, on the other hand, were much better-behaved. When Ag was used as top metal contact a mobility of 1.7 cm^2/Vs was achieved.[20] These results are in agreement with our argumentation and give further evidence that a redox reaction occurs at the interface between DFHCO-4T and easily oxidizable metals, similar to Hirose et al's report that PTCDA reacts with Al, but not with Au and Ag [21].

CONCLUSIONS

DFHCO-4T film growth optimization (high deposition flux and substrate temperature) enables smooth and large grain film morphologies. Using these optimized growth conditions, high performance DFHCO-4T transistors with Au top contacts have been demonstrated. An electron field-effect mobility of 4.6 cm^2/Vs was measured. On the other hand, the mobility of similar transistors fabricated with easily oxidizable top-contact metals is far lower. This reduced transistor performance is attributed to an electron-transfer reaction between the ketone group(s) of DFHCO-4T and metals such as Al or Yb, leading to an insulating interfacial layer. Since Au and Ag – in contrast to Al and Yb – do not react with the semiconducting material, the use of these inert metals is beneficial to achieve high-performance DFHCO-4T-based n-OTFTs.

ACKNOWLEDGMENTS

This work is supported by the EU-funded Project OLAS (contract No. 015034). S. Schols acknowledges the FWO Vlaanderen for financial support.

REFERENCES

1. P. H. Wöbkenberg, J. Ball, D. D. C. Bradley, T. D. Anthopoulos, F. Kooistra, J. C. Hummelen, and D. M. de Leeuw, Appl. Phys. Lett. **92**, 143310 (2008).
2. H. E. Katz, A. J. Lovinger, J. Johnson, C. Kloc, T. Siegrist, W. Li, Y.-Y. Lin, and A. Dodabalapur, Nature 404, 478 (2000).
3. P. R. L. Malenfant, C. D. Dimitrakopoulos, J. D. Gelorme, L. L. Kosbar, T. O. Graham, A. Curioni, and W. Andreoni, Appl. Phys. Lett. 80, 2517 (2002).
4. A. Facchetti, M. Mushrush, M. Yoon, G. R. Hutchison, M. A. Ratner, and T. J. Marks, J. Am. Chem. Soc. 126, 13859 (2004).
5. M. Yoon , C. Kim, A. Facchetti, and T. J. Marks, J. Am. Chem. Soc. **128**, 12851 (2006).
6. D. J. Gundlach, K. P. Pernstich, G. Wilckens, M. Grüter, S. Haas, and B. Batlogg, J. Appl. Phys. **98**, 064502 (2005).
7. T. D. Anthopoulos, B. Singh, N. Marjanovic, N. S. Sariciftci, A. M. Ramil, H. Sitter, M. Cölle, and D. M. de Leeuw, Appl. Phys. Lett. **89**, 213504 (2006).
8. K. Myny, S. De Vusser, S. Steudel, D. Janssen, R. Müller, S. De Jonge, S. Verlaak, J. Genoe, and P. Heremans, Appl. Phys. Lett. **88**, 222103 (2006).
9. L. Chua, J. Zaumseil, J. Chang, E. C. Ou, P. K. Ho, H. Sirringhaus, and R. H. Friend, Nature **434**, 194 (2005).
10. J. Zaumseil and H. Sirringhaus, Chem. Rev. **107**, 1296 (2007).
11. J. A. Venables, G. D. Spiller, and M. Hanbucken, Rep. Prog. Phys. **47**, 399 (1984).
12. M. Yoon , S. A. DiBenedetto, A. Facchetti, and T. J. Marks, J. Am. Chem. Soc. **127**, 1348 (2005).
13. J. Zaumseil, C. L. Donley, J. Kim, R. H. Friend, and H. Sirringhaus, Adv. Mater. **18**, 2708 (2006).
14. J. Grimshaw, *Organic Electrochemistry: An Introduction and a Guide*, 4th edition (Dekker (N.Y.), 2000), chap. 10, Carbonyl Compounds, pp.411-434.
15. H. G. O. Becker, R. Beckert, G. Domschke, E. Fanghänel, W. D. Habicher, P. Metz, D. Pavel, and K. Schwetlick, *Organikum*, 21st edition (Wiley-VCH (Weinheim), 2001), chap. D.7 Reaktionen von Carbonylverbindungen, pp.586-587.
16. M. Hulce and T. Lavaute, Tetrahed. Lett. **29**, 525 (1988).
17. J. C. Kotz and K. F. Purcell, *Chemistry & Chemical Reactivity*, 2nd edition (Saunders College Publishing, 1991), chap. 21 Electrochemistry: The Chemistry of Oxidation-Reduction Reactions, pp.851-899.
18. *CRC Handbook of Chemistry and Physics*, **85**th edition, edited by D. R. Lide (CRC Press (Boca Raton), 2004-2005), pp.8.23-8.33.
19. A. J. Bard and L.R. Faulkner, *Electrochemical Methods: fundamentals and Applications*, 2nd edition (Wiley&Sons, 2001), p. 809.
20. S. Schols, L. Van Willigenburg, R. Müller, D. Bode, M. Debucquoy, S. De Jonge, J. Genoe, P. Heremans, S. Lu and A. Facchetti, Appl. Phys. Lett. **93**, 263303 (2008).
21. Y. Hirose, A. Kahn, V. Aristov, P. Soukiassian, V. Bulovic, S.R. Forrest, Phys. Rev. B, **54**, 13748 (1996).

Mater. Res. Soc. Symp. Proc. Vol. 1154 © 2009 Materials Research Society 1154-B08-04

Field-Induced ESR Spectroscopy on Rubrene Single-Crystal Field-Effect Transistors

Hiroyuki Matsui[1,2] and Tatsuo Hasegawa[1]
[1]Photonics Research Institute (PRI), National Institute of Advanced Industrial Science and Technology (AIST), Tsukuba, Ibaraki 305-8562, Japan
[2]Department of Advanced Materials Science, The University of Tokyo, Kashiwa, Chiba 277-8561, Japan

ABSTRACT

We investigate the electron spin resonance (ESR) spectroscopy for the field-induced carriers in rubrene single-crystal field-effect transistors (SC-FETs), and compare the results with those on pentacene thin-film transistors (TFTs). We observe Lorentz-type ESR signal in rubrene SC-FETs whose linewidth is narrowed with increasing gate voltage and temperature. It demonstrates that the ESR linewidth is determined by motional narrowing effect as we reported on pentacene TFTs. Based on the observations, we discuss the multiple trap-and-release (MTR) processes in the two systems with and without grain boundaries.

INTRODUCTION

Organic field-effect transistors (FETs) have attracted considerable world-wide attentions for their applications toward large-area, low-cost, flexible, and light-weight electronic devices [1,2]. One of the most fundamental issues in organic FETs is to reveal the microscopic nature of charge transport in organic semiconductors, the understanding of which should be key to bringing out the ultimate limit of device characteristics for molecular semiconducting materials [3-5]. Recently, Marumoto and co-workers reported electron spin resonance (ESR) of organic MIS diodes and organic transistors [6,7]. ESR method is quite useful in that it provides a microscopic probe for charge carriers like unpaired electrons and holes [8,9]. Not only to examine the number of carriers and the orientation of molecules, it is possible also to know about carrier dynamics as we discussed in the previous paper [10,11]. By using high-mobility pentacene thin-film transistors (TFTs), we successfully observed the motional narrowing of their ESR spectra; the linewidth of single-Lorentz-type ESR spectra is narrowed as the mobility increases with the variation of gate voltage and temperature. From the analyses of motional narrowing [12], we found that the carrier dynamics is explained by multiple trap-and-release (MTR) model with the average trapping time of a few nanoseconds and the average trap depth of 10 ~ 15 meV [13-15]. However, the origin of the shallow traps and their relation to grain boundaries still remain open questions.

In this paper, we compare the field-induced ESR of single-crystal FETs (SC-FETs), where grain boundaries are not included, with that of pentacene TFTs. We use rubrene as a channel material of SC-FETs because it is known as high-mobility (up to 40 cm^2/Vs) material for single-crystal devices [16,17]. We first show the feature of field-induced ESR at room temperature with a focus on gate-voltage- and angle-dependence. We also show the temperature-dependence of ESR spectra, and point out that the quite similar motional narrowing can be seen in rubrene SC-FETs as in pentacene TFTs. Finally, we discuss the carrier dynamics in the two different systems on the basis of MTR model.

EXPERIMENT

Rubrene SC-FETs were fabricated in top-gate and top-contact geometry, as shown in Figure 1(a). Rubrene single crystals were obtained by physical vapor transport after several purification processes in the same way. We picked up large needle-like rubrene crystals, 12 × 1 mm^2 in size for example, and put silver paste as source and drain electrodes. Gate insulator (parylene C, 1 μm) and gate electrode (Au, 30 nm) were deposited in vacuum. Channel length and width were 10 mm and 1 mm, respectively. The maximum field-effect mobility 0.6 cm^2/Vs was obtained in linear regime of transfer characteristics. The on/off ratio was more than 10^2. Pentacene TFTs were fabricated in the same way we reported before [10]. We show the device structure in Figure 1(b). The maximum field-effect mobility was 0.6 cm^2/Vs, and the on/off ratio was more than 10^4.

ESR measurements were conducted with gate voltage applied (drain voltage was kept to be zero or small enough). *X*-band microwave was mainly used, and *Q*-band microwave was used for low-temperature measurements of rubrene SC-FETs. When we measured angle-dependence of rubrene SC-FETs, we rotated the static magnetic field in the plane perpendicular to the long direction of needle-like crystals.

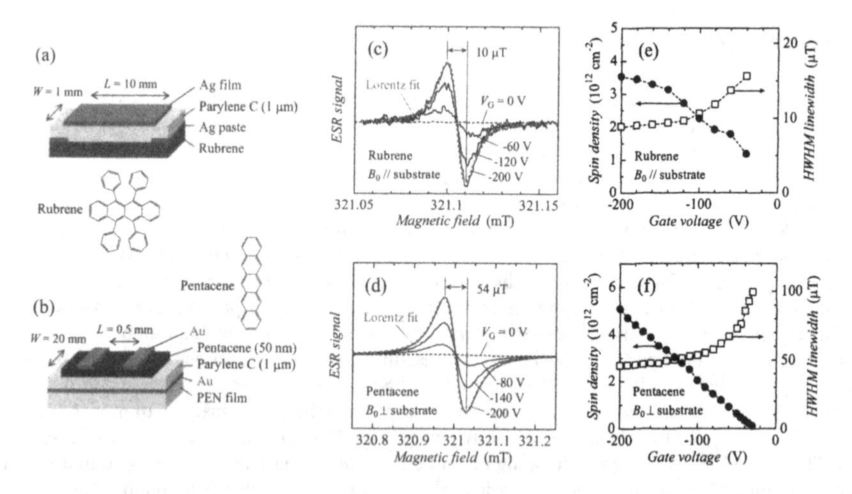

Figure 1. Device structures of (a) rubrene SC-FETs and (b) pentacene TFTs. Field-induced ESR spectra at room temperature for (c) rubrene and (d) pentacene. Spin density and ESR linewidth (half width at half maximum) as functions of gate voltage for (e) rubrene and (f) pentacene.

DISCUSSION

In both kinds of devices, ESR signals were observed at negative gate voltage while no signal was observed at zero or positive gate voltage (Figure 1(c)(d)). The lineshapes can be fitted well by the first derivatives of single Lorentz functions. We calculated the spin density from the second integral of ESR signals, and show it in Figure 2(e)(f) with the ESR linewidth defined as half width at half maximum (HWHM). The obtained spin density agrees well with the simple estimation from gate capacitance. As we discussed in terms of motional narrowing effect previously, the ESR linewidth of pentacene TFTs is narrowed by increasing carrier density. It is explained by the shift of Fermi energy towards valence band edge. Also in case of rubrene SC-FETs, we can see quite similar dependence of ESR linewidth.

Table I shows the anisotropy of *g* value and ESR linewidth. Since ESR measurements see the carriers which are accumulated at semiconductor-insulator interface, this anisotropy reflects the orientation of semiconducting molecules near the interface. Anisotropy of *g* value and linewidth of the pentacene TFTs can be explained by the orientation of pentacene molecules whose long axis is almost perpendicular to the substrate [7,18]. Although the rubrene SC-FETs show relatively small anisotropy, this dependence agrees with the crystal structure of rubrene qualitatively since the *g* value in the direction of $2p_z$ atomic orbitals is expected to be smaller than that in other directions [19].

Table I. Anisotropy of *g* value and ESR linewidth (half width at half maximum).

	Direction of magnetic field B_0	*g* value	Linewidth
Rubrene SC-FETs	B_0 // substrate	2.0023	9.3
	$B_0 \perp$ substrate	2.0024	8.8
Pentacene TFTs	B_0 // substrate	2.0023	31
	$B_0 \perp$ substrate	2.0031	46

ESR linewidth of rubrene SC-FETs show the quite similar temperature dependence to that of pentacene TFTs, as we show in Figure 2. Between 50 K and 200 K, ESR linewidth decreases with increasing temperature for both devices. We explain this feature with motional narrowing effect; carrier motions are activated at high temperature, and make the linewidth narrower by averaging the inhomogeneous local magnetic field. It is noted that we observed Lorentz-type lineshape also in this temperature range. Below 50 K, the carrier motion is so slow that the motional narrowing is not effective any more. The linewidth becomes almost independent of temperature, and the inhomogeneous broadening of hyperfine interactions appears. We see a small upturn of ESR linewidth above 200 K. We confirmed by ESR saturation experiments that this is due to short spin-lattice relaxation time at high temperature.

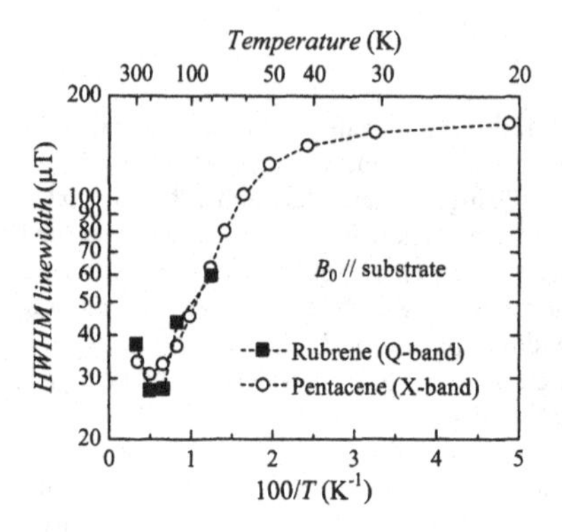

Figure 2. Temperature-dependence of ESR linewidth (half width at half maximum) for rubrene SC-FETs (solid square) and pentacene TFTs (open circle) at V_G = -200 V.

We have now three evidences for motional narrowing effect: Lorentz-type lineshape, narrowing with increasing carrier density, and narrowing with increasing temperature. We can not insist motional narrowing effect only with Lorentz-type lineshape because also ESR spectra determined by spin-lattice relaxation have the same lineshape (we actually see this kind of signal above 200 K). Exchange narrowing can explain the Lorentz-type lineshape and the gate-voltage-dependence, but is inconsistent with the temperature-dependence.

We now discuss the dynamics of charge carriers in the two different systems. The ESR linewidth of rubrene SC-FETs is almost the same as that of pentacene TFTs. Also the magnitude of hyperfine interaction is expected to be in the same order for the two kinds of molecules. Therefore, we can say that the motional frequency of carriers in rubrene SC-FETs is close to that in pentacene TFTs. We note that the activation energy of ESR linewidth in Figure 2 is also similar. Finally, we conclude here that the carrier dynamics in rubrene SC-FETs is quite similar to that of pentacene TFTs, as described by the MTR model with shallow traps. The traps should not be associated with grain boundaries because rubrene SC-FETs do not contain them. We have to consider the trap states inside grains, such as impurities or local molecular displacement. We will report more details about the trap states shortly.

CONCLUSIONS

We compared the ESR spectra of rubrene SC-FETs and pentacene TFTs. We successfully observed motional narrowing effect in rubrene SC-FETs, and found that the results are quite similar to the case of pentacene TFTs. It demonstrates that the traps are not associated with grain boundaries. Even purified single crystals have the shallow traps as seen in polycrystalline pentacene films, which are limiting carrier dynamics in the devices.

ACKNOWLEDGMENTS

This work is supported by Global COE Program "the Physical Sciences Frontier", MEXT, Japan. A part of ESR measurements were done with the facilities in Institute for Molecular Science (IMS), Japan. We thank J. Takeya for supporting the preparation of rubrene single crystals.

REFERENCES

1. C. D. Dimitrakopoulos and D. J. Mascaro, IBM J. Res. Dev. **45**, 11 (2001).
2. A. L. Briseno, R. J. Tseng, M. -M. Ling, E. H. L. Falcao, Y. Yang, F. Wudl, and Z. Bao, Adv. Mater. **18**, 2320 (2006).
3. M. E. Gershenson, V. Podzorov, and A. F. Morpurgo, Rev. Mod. Phys. **78**, 973 (2006).
4. Z. Q. Li, V. Podzorov, N. Sai, M. C. Martin, M. E. Gershenson, M. Di Ventra, and D. N. Basov, Phys. Rev. Lett. **99**, 016403 (2007).
5. A. Troisi and G. Orlandi, Phys. Rev. Lett. **96**, 086601 (2006).
6. K. Marumoto, Y. Muramatsu, Y. Nagano, T. Iwata, S. Ukai, H. Ito, S. Kuroda, Y. Shimoi, and S. Abe, J. Phys. Soc. Jpn. **74**, 3066 (2005).
7. K. Marumoto, S. Kuroda, T. Takenobu, and Y. Iwasa, Phys. Rev. Lett. **97**, 256603 (2006)
8. K. Mizoguchi and S. Kuroda, in *Handbook of Conductive Organic Molecules and Polymers*, edited by H. S. Nalwa (Wiley, Chichester, 1997), vol. 3, pp. 251-317.
9. C. Coulon and R. Clerac, Chem. Rev. **104**, 5655 (2004).
10. H. Matsui, T. Hasegawa, Y. Tokura, M. Hiraoka, and T. Yamada, Phys. Rev. Lett. **100**, 126601 (2008).
11. H. Matsui, and T. Hasegawa, Jpn. J. Appl. Phys. **48**, *in press* (2009).
12. R. Kubo and K. Tomita, J. Phys. Soc. Jpn. **9**, 888 (1954).
13. G. Horowitz and P. Delannoy, J. Appl. Phys. **70**, 469 (1991).
14. G. Horowitz, R. Hajloui, and P. Delannoy, J. Phys. III **5**, 355 (1995).
15. M. F. Calhoun, C. Hsieh, and V. Podzorov, Phys. Rev. Lett. **98**, 096402 (2007).
16. V. Podzorov, V. M. Pudalov, and M. E. Gershenson, Appl. Phys. Lett. **82**, 1739 (2003).
17. J. Takeya, M. Yamagishi, Y. Tominari, R. Hirahara, Y. Nakazawa, T. Nishikawa, T. Kawase, T. Shimoda, and S. Ogawa, Appl. Phys. Lett. **90**, 102120 (2007).
18. C. D. Dimitrakopoulos, A. R. Brown, and A. Pomp, J. Appl. Phys. **80**, 2501 (1996).
19. X. Zeng, D. Zhang, L. Duan, L. Wang, G. Dong, and Y. Qiu, Appl. Surf. Sci. **253**, 6047 (2007).

Photovoltaics

Mater. Res. Soc. Symp. Proc. Vol. 1154 © 2009 Materials Research Society 1154-B09-12

Microstructure and Charge Carrier Transport in Phthalocyanine Based Semiconductor Blends

Andreas Opitz[1], Julia Wagner[1], Bernhard Ecker[1], Ulrich Hörmann[1], Michael Kraus[1], Markus Bronner[1], Wolfgang Brütting[1], Alexander Hinderhofer[2], Frank Schreiber[2]

[1]Institute of Physics, University of Augsburg, Augsburg, Germany

[2]Institute of Applied Physics, University of Tübingen, Tübingen, Germany

ABSTRACT

The continuously growing and wide-spread utilization of blends of organic electron and hole conducting materials comprises ambipolar field-effect transistors as well as organic photovoltaic cells. Structural, optical and electrical properties are investigated in blends and neat films of the electron donor material Cu-phthalocyanine (CuPc) together with fullerene C_{60} and Cu-hexadecafluorophthalocyanine ($F_{16}CuPc$) as electron acceptor materials, respectively. The difference in molecular structure of the spherical C_{60} and the planar molecule CuPc leads to nanophase separation in the blend, causing charge carrier transport which is limited by the successful formation of percolation paths. In contrast, blends of the similar shaped CuPc and $F_{16}CuPc$ molecules entail mixed crystals, as can be clearly seen by X-ray diffraction measurements. We discuss differences of both systems with respect to their microstructure as well as their electrical transport properties.

INTRODUCTION

From the mid 1990s the concept of "bulk-heterojunction solar cells" revolutionized the field of organic photovoltaics: Yu *et al.* reported a polymeric solar cell with an interpenetrating donor/acceptor material system which enables a spatially distributed interface accounting for the small exciton diffusion lengths in organic semiconducting materials [1]. Since that time, blends of organic electron and hole conductive materials are widely used for ambipolar charge carrier transport and photovoltaic cells. The application of distributed interfaces in organic solar cells has the advantage that excitons can efficiently dissociate throughout the whole volume of the organic layer yielding higher amounts of free charge carriers as compared to a bilayer system. Nevertheless, for an efficient transport, each material must provide continuous paths to the contacts. Both aspects entail a competition between efficient charge carrier dissociation and preferably undisturbed transport properties inside the blend.

In this study we present the analysis of two model systems for donor-acceptor blends. These are (i) Cu-phthalocyanine (CuPc) combined with the Buckminster fullerene C_{60} and (ii) CuPc in combination with its fluorinated counterpart $F_{16}CuPc$. While CuPc acts as the donor or p-conductor, C_{60} and $F_{16}CuPc$ are the n-conducting acceptor materials. In addition to studying the fundamental structural and optical properties, centering on the question of phase separation or formation of mixed crystals, we extend our analysis to electrical charge carrier transport properties. The materials used have been previously investigated in similar configurations partially with different donor or acceptor materials [2,3], however, no systematic comparison with respect to the influence of the mixing behavior on the transport properties was reported.

EXPERIMENT

Diodes for ambipolar charge carrier transport and unipolar hole-only devices were fabricated on commercially available indium-tin-oxide (ITO)-coated glass substrates (Merck) which were cleaned with different solvents in an ultrasonic bath followed by an oxygen plasma treatment in order to enhance the work function of ITO and to improve wetability for the aqueous suspension of the intrinsically conducting polymer poly(3,4-ethylenedioxythiophene): poly(styrenesulfonate) (PEDOT:PSS, purchased from H.C. Starck as BAYTRON P). The organic layers were grown by vacuum deposition from low-temperature effusion cells with a base pressure of about 10^{-7} mbar and deposition rates between 0.35 Å/s for neat films and up to 1.4 Å/s for the material with the higher volume fraction in the mixtures.

For the ambipolar setup, a thin (5 Å) layer of LiF was deposited prior to deposition of the Al cathode which was evaporated through a shadow mask to a thickness of 1000 Å, giving an active area of 2×2 mm^2. For hole-only devices, the organic semiconductor is sandwiched between a PEDOT:PSS coated ITO substrate and a Au counterelectrode combined with an electron blocking layer consisting of a 40 nm thick layer of N,N'-bis(3-methylphenyl)-(1,1'-biphenyl)-4,4'-diamine (TPD) or alternatively a thin layer of 2,3,5,6-tetrafluoro-7,7,8,8-tetracyano-quinodimethane (F_4TCNQ). For pure electron transport the organic layer is framed between an Al electrode at the bottom and a LiF/Al counterelectrode on top, restricting the injection to electrons. The overall thickness of the organic layer amounts to 200 nm for the CuPc/C_{60} system and 80 nm for the CuPc/F_{16}CuPc system. The structural formulas of the materials used are depicted in figure 1.

CuPc and F_{16}CuPc were purchased from Sigma Aldrich as sublimed grade and additionally purified by thermal gradient sublimation prior to deposition. C_{60} was purchased from Sigma Aldrich as sublimed grade and used as received. The mixed layers were grown by codeposition from independent evaporation sources, with the deposition rates monitored by two quartz-crystal microbalances.

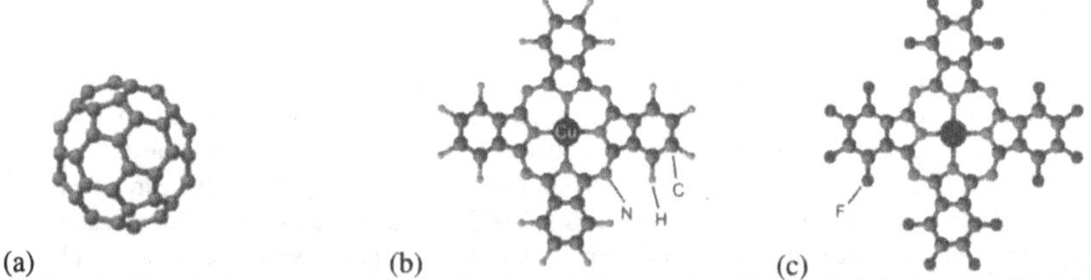

Figure 1. Chemical structure of (a) buckminsterfullerene C_{60}, (b) Cu-phthalocyanine (CuPc) and (c) fluorinated Cu-phthalocyanine (F_{16}CuPc).

Assuming insulating organic films without intrinsic charge carriers and traps, the quantitative analysis of the current-voltage characteristics was realized using the model of trap-free space charge limited currents [4] extended by the Pool-Frenkel like field-dependence of the mobility [5], resulting in a current density given by

$$j_{SCLC} = \frac{9}{8}\varepsilon\mu_0 \frac{V_{eff}^{\ 2}}{d^3} \exp\left(0.89\beta\sqrt{\frac{V_{eff}}{d}}\right),$$

with the zero-field mobility μ_0, the field activation parameter β and the layer thickness d. In order to account for a built-in potential V_{bi} caused by electrodes with different work-functions, an effective voltage V_{eff} is considered which equals the externally applied voltage reduced by V_{bi}. The parameters μ_0, V_{bi} and β are determined by fitting the measured current-voltage characteristics in the higher voltage range.

Optical absorption spectra were measured on neat and blended films deposited on ITO-coated glass substrates covered with PEDOT:PSS using a Varian Cary 50 UV/Vis-spectrophotometer. The X-ray scattering measurements were conducted on a GE/Seifert X-ray diffractometer (Cu Kα1 radiation, multilayer mirror, and double bounce compressor monochromator). While electrical characterization could be realized without exposure to air, X-ray scattering as well as absorption measurements have been performed under ambient atmosphere.

RESULTS and DISCUSSION

Structural properties X-ray scattering measurements, performed in θ-2θ geometry, for neat, mixed and bilayered films are depicted in figure 2 for the material combination CuPc/C_{60} (a,b) and for CuPc/F_{16}CuPc (c). Both types of phthalocyanines show well pronounced diffraction peaks. In addition to previous measurements [6,7], the parallelized and monochromated incident beam enables the detection of the C_{60} diffraction peak.

The diffraction pattern of the CuPc/C_{60} blend displays the same peak positions as detected for the respective neat films and was in the literature assigned to the (200)-reflection of the α-phase of CuPc [8] and the (111)-peak of the fcc structure of C_{60}. This observation rules out the formation of a solid solution but is a clear indicator of phase separation with coexisting crystallites of both materials. This can be ascribed to the different molecular structures of the flat CuPc molecule and the spherical C_{60}. By contrast, the similar molecular structure of CuPc and F_{16}CuPc leads to the formation of a mixed crystal in the blend clearly visible by a diffraction peak which is positioned between the Bragg-reflections of the neat films (see figure 2(c)). This mixed crystal may exhibit a similar structure like the neat phthalocyanines [9,10].

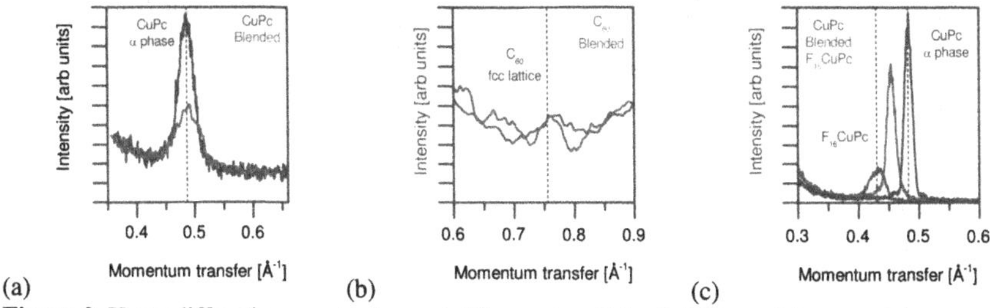

Figure 2. X-ray diffraction spectra for neat, bilayered and blended films of the material system (a,b) CuPc/C_{60} and the (c) CuPc/F_{16}CuPc.

The structural size of the crystallites can be estimated from the width of the diffraction peaks, which is of particular interest in case of phase separated crystals appearing in the CuPc/C_{60} blend. Using the relation $\Delta q_z \sim 2\pi/L$ with Δq_z being the width of the diffraction peak, the coherence volume L corresponding to the size of the crystallites can be evaluated to about

10 nm for both CuPc and C_{60}. Thus, phase separation in this material system can be ascribed to a relatively small length scale in comparison to the large scale phase separation detected in the system pentacene/fullerene [11].

The film morphology of 1:1 blended films of both material combinations are depicted in figure 3, as investigated by non-contact scanning force microscopy (SFM). The materials have been deposited onto SiO_2/Si or PEDOT:PSS/ITO/glass at a substrate temperature of 100 °C. Even at a nominal thickness of 25 nm, maximum heights of up to 58 nm are observed in the blend of C_{60} and CuPc, which approves the model of demixing and phase separation proposed by X-ray scattering. By contrast, the blended CuPc/F_{16}CuPc films show a needle-like structure with similar morphologies as the neat films [7]. This observation confirms the structural result of mixed crystals.

(a) 

(b) 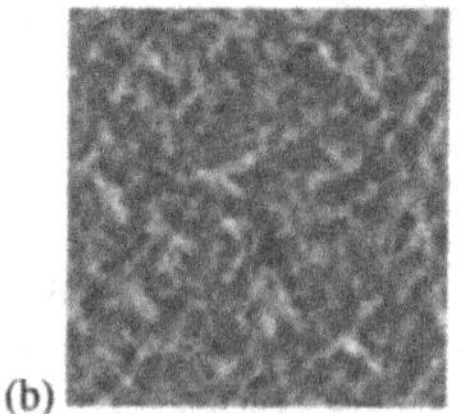

Figure 3. Scanning force microscopy images taken in non-contact mode of 1:1 blended films (nominal thickness of 25 nm) of (a) C_{60} and CuPc deposited on a SiO_2/Si substrate (Max. height: 58 nm, RMS roughness: 6.5 nm) (b) CuPc and F_{16}CuPc deposited on a PEDOT:PSS coated ITO substrate (Max. height: 15 nm, RMS roughness: 2.0 nm) with a total image size of 2×2 μm^2.

Optical absorption spectra Regarding the suitability of the blends for solar cell applications, we analyzed the absorption spectra of neat films as well as blends with a mixing ratio of 1:1 of both material systems, depicted in figure 4. Neat films of CuPc and C_{60} show complementary absorption behavior in the visible spectral range, which qualifies this material combination for application over the whole visible spectrum including the near IR. The spectrum of the blend displays a shape corresponding to a linear combination of both neat spectra, which confirms the model of phase separation [12].

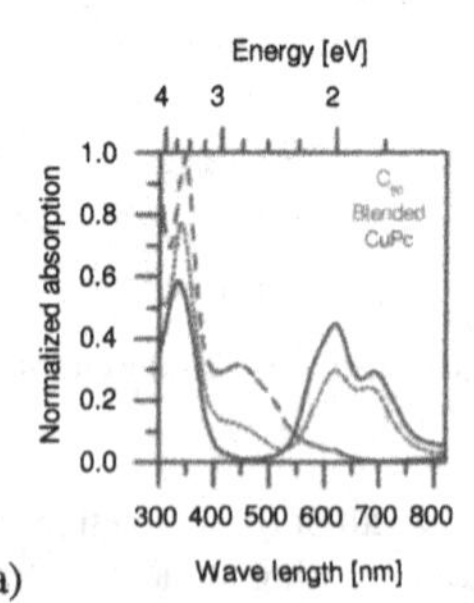

(a)

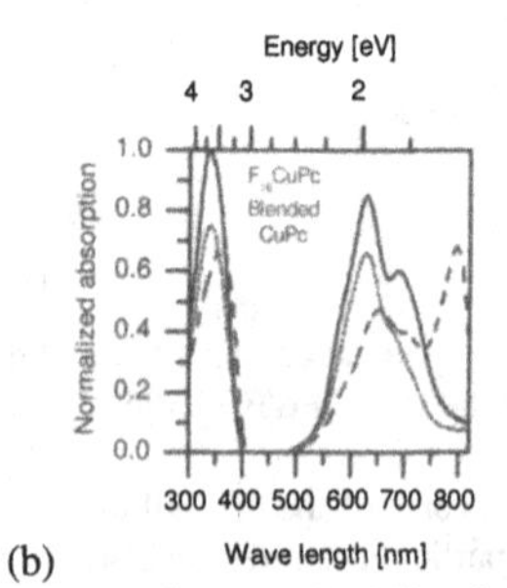

(b)

Figure 4. Absorption spectra in the UV-Vis range of neat films and 1:1 blends: (a) CuPc/C_{60}; (b) CuPc/F_{16}CuPc.

In contrast, both CuPc and F_{16}CuPc show hardly any absorption in the blue-green wavelength range of the visible spectrum between the two absorption edges at $\lambda = 400$ nm and 550 nm, making this material combination not very favorable for sunlight application. The slightly differing curves in the higher wavelength range can be assigned to the difference in the interaction of the molecules in the unit cell [3]. When mixing both molecules the intensity of the peak at 793 nm, which may be attributed to a non-herringbone structure [10], is reduced drastically.

Electrical properties Investigations of electrical transport properties of neat and blended films have been realized by fabricating hole-only, electron-only and ambipolar diodes. The corresponding *I-V* characteristics are published elsewhere [13,14] and skipped here for brevity. They were analyzed using the trap-free SCLC model described above. The obtained zero-field mobilities of the CuPc/C_{60} material system in dependence on the concentration are summarized in figure 5(a). It can be seen that both charge carrier types are transported in CuPc as well as in C_{60}, even though the unipolar mobilities depend strongly on the mixing ratio. Starting from neat C_{60} the electron mobility decreases exponentially with increasing CuPc addition and shows a further reduction when switching over to neat CuPc. These aspects allow for the conclusion that the electron transport in the blend is predominantly carried by C_{60} molecules and decreases in the blend where the hopping distance between the molecules is increased. The unipolar hole mobility changes uniformly over the whole concentration range. As a consequence of the ambipolar nature of both CuPc and C_{60}, the ambipolar mobility in the blends is higher than the sum of unipolar hole and electron mobility. A consolidated view of these factors indicates that the charge carrier transport in blended films is mainly based on the excellent electron transport properties inside conductive paths of C_{60}.

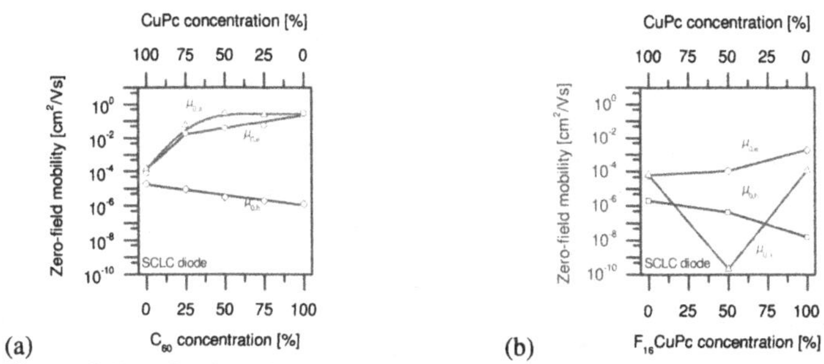

Figure 5. Zero-field mobilities for electron-only, hole-only, and ambipolar transport determined from the SCLC model for neat and blended films of CuPc/C_{60} (a) and CuPc/F_{16}CuPc (b).

The zero-field mobilities of the CuPc/F_{16}CuPc material system are depicted in figure 5(b). The unipolar mobilities in the blended films are in-between the ones of the neat films, indicating that the blend contains mixed crystals with similar π-π overlaps as compared to the neat films. In contrast to the CuPc/C_{60} material system, the ambipolar mobility inside the CuPc/F_{16}CuPc blend is several orders of magnitude lower than both unipolar mobilities as well as the ambipolar mobilities of the neat films. Since the unipolar mobilities are high, it is hardly probable that the extremely low ambipolar transport in the blend is due to an absence of percolation paths. Instead, the reduced ambipolar mobility is probably related to the simultaneous presence of both charge carrier types. A tentative explanation may be found in the generation of charge transfer (CT) excitons which might be created by the injection of both charge carrier types [15]. As a result, these CT excitons would limit the transport by blocking the occupied molecules for further injected charge carriers. The generation of CT excitons can be facilitated by the high electron affinity of the F_{16}CuPc which is located close to the ionization potential of CuPc [16]. For proving this phenomenon further work is in progress.

SUMMARY

It has been demonstrated that blends of CuPc and C_{60} as compared to mixed films of CuPc and F_{16}CuPc display different types of film growth, namely phase separation and mixed crystal formation. The transport properties of the phase separated blend of CuPc/C_{60} are based on percolation paths of the different phases and dominated by the electron transport in C_{60}. By contrast, blends of the phthalocyanines CuPc and F_{16}CuPc demonstrate the formation of mixed crystals. A drastically reduced ambipolar mobility was found in the blend and might be assigned to the generation of charge transfer excitons in neighboring CuPc and F_{16}CuPc molecules created by the simultaneous injection or photogeneration of both charge carrier types.

ACKNOWLEDGMENTS

This work was supported by the Deutsche Forschungsgemeinschaft through priority programs 1121 and 1355. The authors thank Jens Pflaum (Universities of Stuttgart and Würzburg) for purifying organic materials.

REFERENCES

1. G. Yu *et al., Science* **270**, 1789-1791 (1995).
2. P. Peumans *et al., Nature* **425**, 158–162 (2003).
3. J. O. Ossó *et al., Adv. Func. Mater.* **12**, 455-460 (2002).
4. N. Mott and R. Gurney, Electronic Processes in Ionic Crystals (Clarendon Press, Oxford, 1940).
5. P. Murgatroyd, *J. Phys. D: Appl. Phys.* **3**, 151–156 (1970).
6. B. P. Rand *et al., J. Appl. Phys.* **98**, 124902 (2005).
7. M. Bronner *et al., phys. stat. sol.* (a) **205**, 549–563 (2008).
8. O. Berger *et al., J. Mater. Sci.- Mater. El.* **11**, 331-346 (2000).
9. Z. Bao *et al., J. Am. Chem. Soc.* **120**, 207–208 (1998).
10. D. G. de Oteyza *et al., J. Am. Chem. Soc.* **128**, 15052–15053 (2006).
11. I. Salzmann *et al., J. Appl. Phys.* **104**, 114518 (2008).
12. D. Datta *et al., Thin solid films* **516**, 7237–7240 (2008).
13. A. Opitz *et al., SPIE Proc.* 7002, 70020J (2008)
14. A. Opitz *et al., Org. Electron.* (2009) submitted.
15. M. Pope and C. E. Swenberg, Electronic processes in organic crystals and polymers (Oxford University Press, New York, 1999)
16. M. Knupfer and H. Peisert, *phys. stat. sol.* (a) **201**, 1055-1074 (2004).

Poster Session II

Mater. Res. Soc. Symp. Proc. Vol. 1154 © 2009 Materials Research Society 1154-B10-09

Tuning the Threshold Voltage in Organic Field Effect Transistors by Space Charge Polarization of Gate Dielectric

Heisuke Sakai, Koudai Konno and Hideyuki Murata
School of Materials Science, Japan Advanced Institute of Science and Technology, Asahidai, Ishikawa 923-1292, Japan

ABSTRACT

We demonstrate a tunable threshold voltage in an organic field effect transistor (OFET) using ion-dispersed gate dielectric. Application of external electric field (V_{ex}) to the gate dielectrics causes the dispersed ions in the gate dielectric to migrate by electrophoresis and form space charge polarization. The threshold voltage (V_{th}) decreases from -11.3 V to -6.1 V. The shift direction of V_{th} is easily tuned by the polarity of the external voltage. These shifts are attributed to the ion migration in the ion-dispersed gate dielectric. The UV-VIS differential absorption spectra of the OFETs indicate that the active layer is doped by the migrated anions. This result indicates the active layer was charged not only electrostatically but also electrochemically in OFET without buffer layer. By inserting a buffer layer between the active layer and the ion-dispersed dielectric, the reaction was effectively prevented.

INTRODUCTION

In recent years, organic field-effect transistors (OFETs) using organic semiconductors have attracted much research interest due to their unique advantages, which include a variety of molecular designs, light weight, low cost of fabrication and mechanical flexibility. In particular, research interest into gate dielectrics is recently increasing, since the choice of the gate dielectrics directly affects the electric properties of OFETs [1,2].

Recently we have reported the effect of gate dielectric polarization on the electric characteristics of OFETs, where the polarization was achieved by the alignment of permanent dipoles in polymer chains [3]. The polarization of gate dielectrics induced the mobile charge carrier in the semiconductor layer at the semiconductor-dielectric interface and as a result, V_{th} shifted toward a lower voltage. By using polarized gate dielectrics, V_{th} of OFETs was decreased by 6 V compared to that of OFETs using non-polarized gate dielectrics. However, the decrease of V_{th} was limited by the small amount of polarization, a limitation which was caused by the difficulty of aligning permanent dipoles in the solid dielectric film. For a further decrease of V_{th}, novel approaches will be necessary to realize a higher magnitude of polarization in the gate dielectrics.

Ion migration in polymer gate dielectric is one of the candidates to realize a higher magnitude of polarization of gate dielectric, because ions in a matrix polymer can easily migrate due to the influence of an external electric field and form a space charge polarization [4,5]. This polarization is easily tunable by using the number of dispersed ions and the ion migration distance. The ion migration distance can be controlled by the time an external electric field is applied. In this study, we report that the reduction of V_{th} can be achieved by using the ion-dispersed gate dielectric. Results of ultraviolet/visible (UV-VIS) absorption study reveal that the active layer of OFETs is charged not only electrostatically but electrochemically with increasing the time for application of V_{ex}.

EXPERIMENT

We prepared two types of gate dielectric for this study. One consists of ion-dispersed polymer (denoted as OFET without buffer layer), and the other consists of double layer of ion-dispersed polymer and polymer buffer layer (denoted as OFET with buffer layer). Polymethylmethacryrate (PMMA) and 10-methyl-9-phenylacridinium perchlorate ($MPA^+ClO_4^-$) were dissolved in acetonitrile, where the mol ratio of constitutional repeating unit of PMMA to ionic compound was 50 to 1. The dielectric film was fabricated by spin-coating on ITO/glass substrate, and the film thickness was 340 nm. To prepare the OFET with buffer layer, the PMMA buffer layer was fabricated by spin-coating of the PMMA xylene solution (10 wt %) on the ion-dispersed layer film after annealing the ion-dispersed layer at 100 °C for 2 h in air. The role of PMMA buffer layer is to delay the migration of anion when an external electric field is applied for 10 min, and it prevents the chemical doping of MPA^+ClO4^- into the active layer. The total thickness of the ion-dispersed layer and the buffer layer was 690 nm. ITO was used as the gate electrode in our devices. As an active layer of OFETs, thin films of pentacene (Aldrich, purified by vacuum sublimation) were formed by vacuum deposition at the rate of 0.03 nm s^{-1} on the dielectrics, followed by the deposition of a gold comb-shaped electrode at the a deposition rate of 0.03 nm s^{-1} through a shadow mask. The film thicknesses of pentacene and gold were 30 nm and 50 nm. The channel length (L) and channel width (W) were 75 μm and 24500 μm. Electric characteristics of OFETs were measured with a Keithley 4200 semiconductor characterization system in dry nitrogen atmosphere. Capacitances of the dielectrics were measured with a capacitor structure of glass substrate /ITO electrode/ion-dispersed dielectrics/Al electrode using an Agilent 4284A LCR meter.

When we applied V_{ex} to the gate electrode with a Keithley 2400 SourceMeter for a certain bias time (T_{bias}), both source and drain electrodes on the gate dielectrics were grounded and used as counter electrodes. While V_{ex} is applied to the gate dielectrics, the ions of MPA^+ and ClO_4^- migrate in opposite directions and form space charge polarization in the gate dielectric. Comparing the sizes of MPA^+ cations (10 Å) and ClO_4^- anions (3.5 Å) with the diameter of the free volume hole of PMMA (5.4 Å) [6], the cations can hardly migrate in the dielectric due to the molecular sieve effect of PMMA polymer chains. Also, MPA^+ cations would be stabilized by the interaction with oxygen in a carbonyl group of PMMA. Therefore, the polarization of our dielectrics would be mainly due to the migration of ClO_4^- anions. The size of the MPA^+ cation and of the $ClO4^-$ anion was calculated with a density functional theory (DFT) method with $DMol^3$ developed by Accelrys Inc. The geometry optimization was performed and the structural properties were obtained at the Generalized Gradient Approximation (GGA) level with the BLYP/DNP method.

UV-VIS differential absorption spectra of the OFETs were measured with a JASCO V-570 spectrometer. To separate the $MPA^+ClO_4^-$ ion pairs in the dielectrics, we applied V_{ex} to ITO electrode for T_{bias}. The differential absorption spectra were measured by subtracting the absorption spectra without the electric field from those after applying the electric field for a certain bias time.

DISCUSSION

Figures 1a and 1b show the output characteristics of the OFET without buffer layer at the T_{bias} of 0 min and 10 min. The gate voltages were changed from 5 V to -30 V in steps of -5 V. The output characteristics clearly show that drain current (I_D) increases after V_{ex} was applied, where I_D value at V_G = -30 V with T_{bias} = 10 min (Fig. 1b) increases over 1.3 times compared to that with T_{bias} = 0 min (Fig. 1a). This result indicates that the number of the charge carriers was increased after applying V_{ex}. Increasing I_D with increasing T_{bias} indicates that the magnitude of polarization of the gate dielectrics was enhanced by the separation of $MPA^+ClO_4^-$ ion pairs, and as a result the active layer was charged.

Figure 2a shows the change in transfer characteristics of the OFET without buffer layer at V_D=-30 V, measured at different T_{bias} (V_{ex}=-30 V) from 0 to 10 min with a 2 min period. At T_{bias}=0 min, we measured the transfer characteristics twice and observed identical transfer characteristics. This confirms that the measurement does not affect the transfer characteristics, and that the shift in the transfer characteristics in Fig. 2a is mainly caused by applying V_{ex}. As shown Fig 2a, the "off current" of the device increased with increasing T_{bias}. In the case of OFET without buffer layer, it is possible that the ions which interacted electrochemically with the active layer and enhanced the I_D [7,8]. That is, the ClO_4^- anions penetrated into the semiconductor layer and doped the pentacene. As a result, doped pentacene layer showed high conductivity [9] and off current of OFET without buffer layer increased. In contrast, the "off current" of the OFET with buffer layer did not increase with increasing T_{bias} [10]. These results suggest that the buffer layer effectively prevented interaction between the active layer and the ClO_4^- anions. To obtain further insight into the interaction, we confirmed the electrochemical interaction at the active layer with UV-VIS reflection absorption spectra of the OFETs *(vide infra)*.

V_{th} was estimated through a linear fit and its intercept ($I_D^{1/2}$ = 0) of the plot of the square root of I_D versus V_G, As shown in Fig 2b, the V_{th} decreased from -11.3 V (T_{bias}=0 min) to -6.1 V (T_{bias}=10 min). After the measurement of the V_{th} with negative bias (V_{ex} = -30 V), applying the V_{th} with the positive bias (V_{ex} = 30 V) enabled return to the initial V_{th} (data not shown). On the other hand, in the OFET with the buffer layer, we applied a higher V_{ex} (60 V) than was applied in the OFET without buffer layer, because the total thickness of gate dielectric became about twice by inserting the buffer layer. V_{th} was decreased from -35.1 V (T_{bias}=0 min) to -13.1

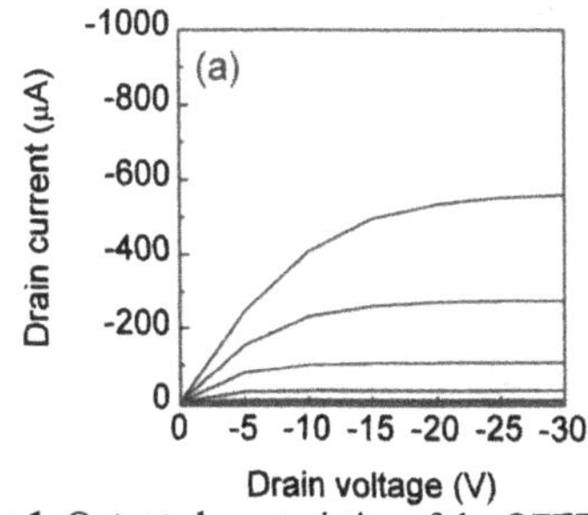

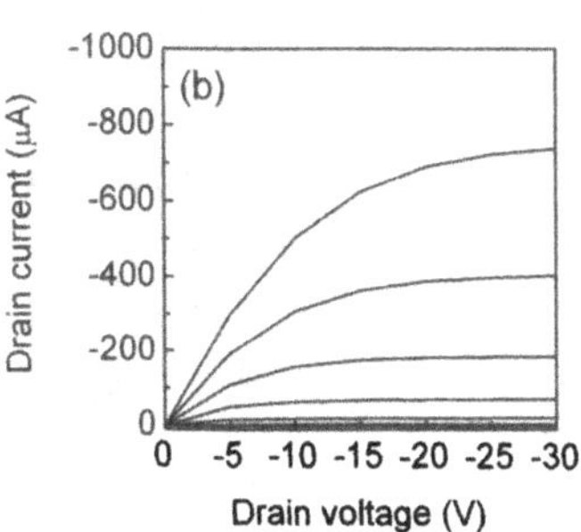

Figure 1. Output characteristics of the OFET by external electric field of V_{ex} = -30 V at (a) T_{bias} = 0 min and (b) T_{bias} = 10 min. The gate voltages were changed from 5 V to -30 V in steps of -5 V.

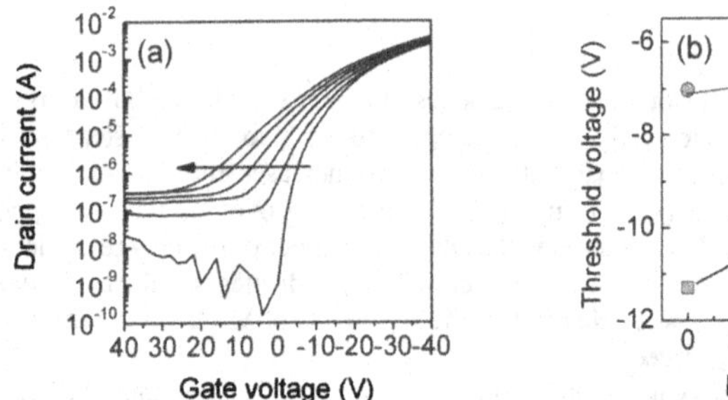

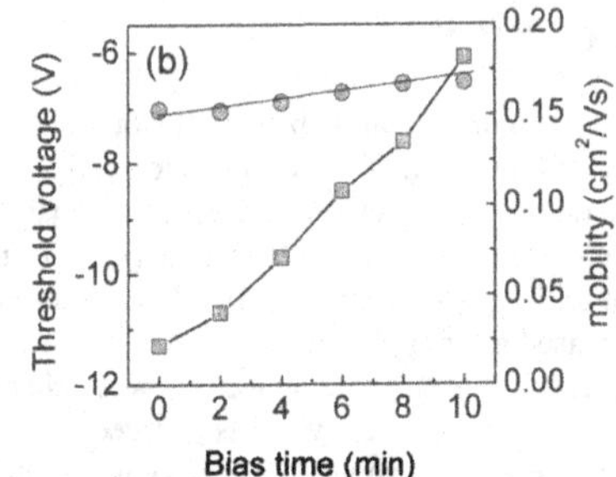

Figure 2. (a)Transfer characteristics (V_D = -30 V) for the OFETs with ion-dispersed gate dielectrics depending on various T_{bias}. T_{bias} varies from 0 to 10 min in steps of 2 min. An arrow shows the shift direction of the transfer curve with increasing T_{bias}. (b) Changes in V_{th} (square) and of the mobility (circle) as a function of T_{bias}.

V (T_{bias}=10 min) by applying the V_{ex}. A large shift of 22 V was achieved. By applying V_{ex} to OFET with buffer layer, the $ClO4^-$ anions migrate into the PMMA buffer layer, and as a result, the ion migration distance of OFET with buffer layer can be longer than that of OFET without buffer layer. Thus, the magnitude of the polarization became larger, and the V_{th} shift of OFET with buffer layer was larger than that of OFET without buffer layer. The detailed data of OFET with buffer layer were reported elsewhere [10].

The carrier mobility was calculated by fitting the plot of the square root of I_D versus V_G [11]. As shown in Fig 2b, the mobility of the OFET without buffer layer increased slightly with increasing T_{bias}. The capacitance measurement of the ion-dispersed gate dielectrics revealed that the dielectric permittivity of the gate dielectrics did not change after T_{bias} was applied. In the other hand, as the active layer was doped with increasing T_{bias}, the number of carriers in the active layer increased, and as a result, the drain current increased. This current reflects as the increase of mobility with increasing T_{bias}.

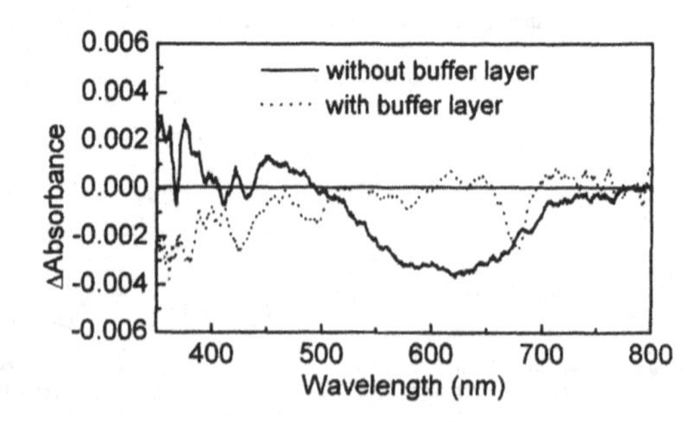

Figure 3. UV-VIS differential absorption spectra of the OFETs without buffer layer (solid line) and with buffer layer (dotted line) after application of V_{ex} (-60 V) for 10 min.

Figure 3 shows the UV-VIS differential absorption spectra of the OFETs without and with buffer layer. These spectra were obtained by subtracting the spectra at V_{ex} = 0 V from those applied at V_{ex} = -60 V for T_{bias} = 10 min. Minakata et al. reported the absorption band at around 400 - 500 nm increased due to the formation of pentacene cation, and the absorption band at around 500 - 700 nm decreased due to the decrease of the amount of neutral pentacene [9]. The broad absorption at around 400 - 500 nm increased in the OFETs without buffer layer. At the same time, the absorption at 500 - 700 nm significantly decreased. These results suggest that the pristine pentacene was doped after V_{ex} was applied. Therefore, we concluded that the active layer was charged not only electrostatically but also electrochemically in the OFET without buffer layer. That is, the origins of the Vth shift of the OFET were both electrostatic effect and electrochemical effect. In the OFET with the buffer layer, the increase of absorption intensity around 400-500 nm and the bleaching at 500 – 700 nm were not observed. These results clearly show that the electrochemical doping in the active layer was effectively prevented by the buffer layer. Thus, the active layer of the OFET was only charged electrostatically.

CONCLUSIONS

We have demonstrated that OFETs using ion-dispersed gate dielectric were significantly affected by ion migration. Transfer curve shifted lower with increasing T_{bias} indicating that the V_{th} of the device decreased by applying V_{ex}. The V_{th} shift of 5.2 V was readily achieved increasing T_{bias}. The increase of "off current" in the transfer characteristics suggests that the active layer was doped by $ClO4^-$ anions. This result was confirmed by UV-VIS differential absorption spectra. Thus, we concluded that the active layer was charged not only electrostatically but also electrochemically in the OFET without buffer layer. In addition, the reaction was effectively prevented by inserting a buffer layer between the active layer and the ion-dispersed dielectric.

ACKNOWLEDGMENTS

H. S. gratefully acknowledges financial support by JSPS Research Fellowships for Young Scientists (doctoral course).

REFERENCES

1. J. Veres, S. Ogier, G. Lloyd, D. deLeeuw, *Chem. Mater.* **16**, 4543 (2004).
2. A. Facchetti, M.-H. Yoon, and T. J. Marks, *Adv. Mater.* **17**, 1705 (2005).
3. H. Sakai, Y. Takahashi, H. Murata, *Appl. Phys. Lett.* **91**, 113502 (2007).
4. J. R. Macdonald, J. Chem. Phys., 58, 4982, (1973).
5. E. H. Snow and M. E. Dumesnil, J. Appl. Phys. 37, 2123 (1966).
6. K. Suvegh, M. Klapper, A. Domjan, S. Mullins, W. Wunderlich, A. Vertes, *Macromolecules* **32**, 1147 (1999).
7. M. J. Panzer and C. D. Frisbie, *J. Am. Chem. Soc.* **129**, 6599 (2007).
8. J. D. Yuen, A. S. Dhoot, E. B. Namdas, N. E. Coates, M. Heeney, I. McCulloch, D. Moses, and A. J. Heeger, *J. Am. Chem. Soc.* **129**, 14367 (2007).
9. T. Minakata, I. Nagoya, and M. Ozaki, *J. Appl. Phys.* **69**, 7354 (1991).

10. H. Sakai, K. Konno, H. Murata, *Appl. Phys. Lett.* **94**, 073304 (2009).
11. S.M. Sze, "*Physics of Semiconductor Devices*", (Wiley, 1981).

Mater. Res. Soc. Symp. Proc. Vol. 1154 © 2009 Materials Research Society 1154-B10-10

Spin Transport and Magneto-Resistance in Organic Semiconductors

Mohammad Yunus and P. Paul Ruden
University of Minnesota, Minneapolis, Minnesota 55455
Darryl L. Smith
Los Alamos National Laboratory, Los Alamos, New Mexico 87545

ABSTRACT

Calculated results for spin injection, transport, and magneto-resistance (MR) in organic semiconductors sandwiched between two ferromagnetic contacts are presented. The carrier transport is modeled by spin dependent device equations in drift-diffusion approximation. In agreement with earlier results, spin injection from ferromagnetic contacts into organic semiconductors can be greatly enhanced if (spin-selective) tunneling is the limiting process for carrier injection. Modeling the tunnel processes with linear contact resistances yields spin currents and MR that tend to increase with increasing bias. We also explore the possibility of bias dependent contact resistances and show that this effect may limit MR to low bias.

INTRODUCTION

π-conjugated organic semiconductors are materials of choice for optoelectronic and photovoltaic devices due to their potential for low-cost and large-area fabrication. Displays based on organic light emitting diodes (OLEDs) are already seeing commercial use.[1] So-called spintronic devices have shown considerable potential for a great extension of device performance and functionality.[2] Commercial successes of metal-based spintronic devices have been achieved in magnetic recording heads and memories that use the giant magnetoresistance (GMR) and tunneling magnetoresistance (TMR) effects.[3] Intense research efforts are now focused on extending spintronics into the realm of semiconductor devices. However, spin injection from a ferromagnetic (FM) contact into a semiconductor is a challenging task.[4] A tunnel barrier with spin selective transmission probability between the FM contact and the semiconductor can greatly enhanced spin injection[5,6,7,8] and there has been some success with spin injection and detection using inorganic semiconductors.[9] Organic semiconductors also appear to be promising materials for spin transport. The weak spin-orbit and hyperfine interaction make the spin coherence time long,[10] and the fabrication of organic semiconductors with strongly spin-polarized $La_{0.7}Sr_{0.3}MnO_3$ (LSMO) contacts provides additional potential for organic semiconductor spintronics. LSMO is to form tunnel contacts to organic semiconductors with a suitable contact resistance.[11]

Spin injection due to a tunnel barrier at the injecting contact by itself does not cause measurable MR. Effective spin valves need also a similarly spin-selective collecting contact. MR effects in organic spin valve structures have been reported in the literature.[12,13] In some cases these devices have FM contacts made from LSMO. Since the organic semiconductor layer thicknesses are much larger than reasonable tunnel lengths, carrier transport in the semiconductor is expected to be diffusive, and the observed MR is not attributed to tunneling from one metal contact to the other (TMR). In this paper we explore further the effects of spin-selective contact resistances on spin injection and MR.

THEORY

For convenience, the model is formulated for negative charge carriers (electrons). In many cases, the current in the organic semiconductor may in fact be due to positive charge carriers (holes), but this does not affect the results and conclusions reached. The organic semiconductor is characterized by the electron mobility, μ_n, a spin relaxation time constant, τ_s, and an effective density of states for the conduction band, n_0, which is approximately equal to the molecular (or monomer) density. Steady-state carrier transport in the organic semiconductor is governed by the spin-dependent continuity equations coupled with Poisson's equation.

$$\mu_n \frac{d}{dx}\left(n_\uparrow E + \frac{kT}{e}\frac{dn_\uparrow}{dx}\right) - \frac{n_\uparrow - n_\downarrow}{\tau_s} = 0, \tag{1a}$$

$$\mu_n \frac{d}{dx}\left(n_\downarrow E + \frac{kT}{e}\frac{dn_\downarrow}{dx}\right) - \frac{n_\downarrow - n_\uparrow}{\tau_s} = 0, \tag{1b}$$

$$\frac{dE}{dx} = -e(n_\uparrow + n_\downarrow)/\varepsilon \tag{2}$$

where the spin-up (SU) and spin-down (SD) currents are expressed in drift-diffusion approximation. Here $n_{\uparrow,\downarrow}$ are the spin-dependent electron densities, E is the electric field, kT is the thermal energy, and ε is the permittivity of the semiconductor. Non-degenerate carrier statistics is assumed for the organic semiconductor. An analytical solution for the injected spin and charge currents is obtainable under the following approximations, (i) τ_s is long compared to the transit time, i.e. the last terms in eqs. 1 can be neglected; (ii) the electric field is constant throughout the organic semiconductor; and (iii) ohmic boundary conditions apply, i.e. the quasi-Fermi levels are continuous at the contacts. The resulting charge current, J, and spin current, J_S, in the organic semiconductor can then be expressed as,

$$J = F(V')\left[\exp\left(\frac{V'}{kT}\right) \times \cosh\left(\frac{\Delta\mu(0)}{2kT}\right) - \cosh\left(\frac{\Delta\mu(d)}{2kT}\right)\right] \tag{3a}$$

$$J_S = F(V')\left[\exp\left(\frac{V'}{kT}\right) \times \sinh\left(\frac{\Delta\mu(0)}{2kT}\right) - \sinh\left(\frac{\Delta\mu(d)}{2kT}\right)\right] \tag{3b}$$

V' is the voltage dropped across the semiconductor, and $F(V')$ is given by,

$$F(V') = \frac{-e\mu_n \frac{V'}{d} n_0 \exp\left(-\frac{\Phi_B}{kT}\right)}{\exp\left(\frac{V'}{kT}\right) - 1} \tag{4}$$

Here Φ_B is the Schottky barrier height (symmetric for both contacts) and d is the thickness of the device. $\Delta\mu$ (0) and $\Delta\mu$ (d) are the difference of the quasi-Fermi levels ($\mu_\uparrow$ - $\mu_\downarrow$) for the SU and

SD electrons at the left and right contacts respectively. We assume $\tau_s >> d^2/(\mu_n kT/e)$, hence the injected spin current density is constant throughout the semiconductor. The difference of the quasi-Fermi levels at the left and right contacts are related to the spin-dependent resistances of the tunnel barrier and are given by[14]

$$\Delta\mu(0) = +\frac{1}{2}e(R_{\uparrow L} - R_{\downarrow L})J + \frac{1}{2}e(R_{\uparrow L} + R_{\downarrow L})J_S(0) \tag{5a}$$

$$\Delta\mu(d) = -\frac{1}{2}e(R_{\uparrow R} - R_{\downarrow R})J - \frac{1}{2}e(R_{\uparrow R} + R_{\downarrow R})J_S(d) \tag{5b}$$

$R_{\uparrow,\downarrow;L,R}$ are the spin-dependent (tunnel) contact resistances for the SU and SD electrons at the left (L) and right (R) electrodes. We also assume that there is no spin scattering. Equations 3 and 5 can be solved self-consistently for a particular bias, V'. The total voltage applied to the device is then obtained from,

$$V = V' + \Delta V_L + \Delta V_R \tag{6a}$$

$$\Delta V_{L,R} = -(1/4)[(R_{\uparrow L,R} + R_{\downarrow L,R})J + (R_{\uparrow L,R} - R_{\downarrow L,R})J_S] \tag{6b}$$

We assume that the polarization of the left (injecting) electrode is always in the up-direction. For parallel (P) alignment of contact magnetizations, the polarization of the right (extracting) electrode is in the up-direction, whereas for anti-parallel (AP) alignment, the polarization of the right electrode is in the down direction. The difference between parallel (P) and anti-parallel (AP) alignments of the contact magnetizations is expressed through the terms involving $R_{\uparrow R} - R_{\downarrow R}$, which have the same sign as $R_{\uparrow L} - R_{\downarrow L}$ in P configuration, but opposite sign in AP configuration. Thus the currents and total applied voltages for the two contact alignments are obtained as a function of V'. Finally, the magneto-resistance is defined as,

$$\mathrm{MR}(\%) = \left(\frac{V_{AP}}{V_P}\frac{J_P}{J_{AP}} - 1\right) \times 100 \tag{7}$$

Here V_{AP} and V_P denote the applied biases for the P and AP configurations, and J_P and J_{AP} are the corresponding charge current densities.

RESULTS AND DISCUSSION

For all calculations, the Schottky barrier height, Φ_B is 0.2eV, the electron mobility, μ_n is 10^{-2} cm^2/Vsec and the effective density of states, n_0 is 10^{21}cm^{-3}. The device thickness is 100nm. The spin relaxation time thus needs to be much greater than 0.4μsec to be consistent with our approximation above. Figure 1 shows the calculated spin polarizations J_S/J for different combinations of contact resistances and figure 2 shows the corresponding MRs. The values of the contact resistances $R_\uparrow$ and $R_\downarrow$ are $4\times10^{-3}\Omega$cm^2 and $20\times10^{-3}\Omega$cm^2 for the solid lines, $4\times10^{-3}\Omega$cm^2 and $8\times10^{-3}\Omega$cm^2 for the dashed lines, and $2\times10^{-2}\Omega$cm^2 and $4\times10^{-2}\Omega$cm^2 for the dotted lines, respectively. The values of the contact resistances are chosen to compare the effects of different $R_\downarrow/R_\uparrow$ ratios and of different magnitudes. At low bias, both contacts control the spin current in the device, consequently the current polarization for the P and AP alignments of the contacts are different. At high biases, only the injecting contact exercises control and the current polarizations for the two configurations become nearly equal. Spin injection and MR increase strongly (solid lines) or moderately (dashed and dotted lines) with applied bias. At low bias, spin injection can be increased either by increasing $R_\uparrow/R_\downarrow$ (solid lines compared to the other two cases) or by increasing the magnitude of the contact resistances (dotted lines compared to the dashed lines). However, at high bias, only the ratio $R_\uparrow/R_\downarrow$ (of the injecting contact) matters, hence the dashed lines and the dotted lines tend to converge.

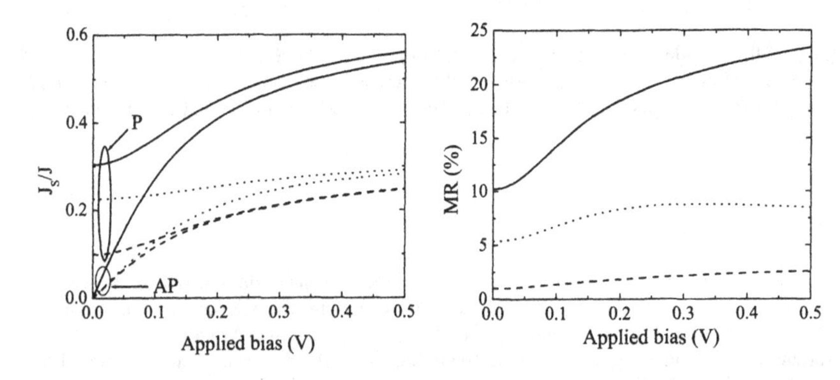

Figure 1. Spin polarization, J_S/J for P and AP contact magnetization alignments. The $R_\uparrow/R_\downarrow$ ratios are 5 (solid lines) and 2 (dashed and dotted lines, where the magnitudes are different).

Figure 2. MR plotted as a function of the applied bias for three cases of different contact resistances shown in figure 1

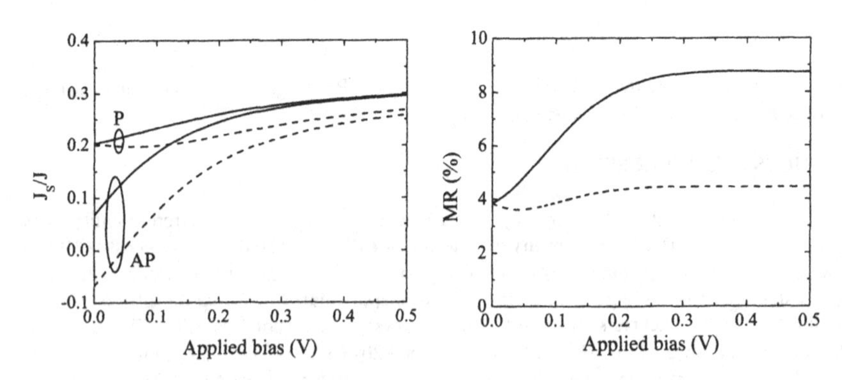

Figure 3. Spin polarization, J_S/J for P and AP configurations for non-symmetric contact resistances. ($R_\uparrow / R_\downarrow = 2$).

Figure 4. Corresponding MR for the cases shown in figure 3.

Non-symmetric contact resistances are considered in figures 3 and 4. The contact resistances are $R_{\uparrow L} = 2\times10^{-2}\Omega\text{cm}^2 = 2R_{\uparrow R}$, $R_{\downarrow L} = 4\times10^{-2}\Omega\text{cm}^2 = 2R_{\downarrow R}$ for the solid lines, and $2R_{\uparrow L} = 2\times10^{-2}\Omega\text{cm}^2 = R_{\uparrow R}$, $2R_{\downarrow L} = 4\times10^{-2}\Omega\text{cm}^2 = R_{\downarrow R}$ for the dashed lines. When $R_L > R_R$, the left contact controls spin injection at low bias, hence the current polarization is positive for both P and AP configurations. On the other hand, the right contact dominates at low bias when $R_L < R_R$ and the polarization is positive for the P configuration but negative for the AP configuration. Of course, the MR is the same at zero bias for both cases.

The model calculations above yield magneto-resistances that increase strongly or at least moderately with applied bias. Most experiments to date appear to show decreasing MR with increasing bias.[12] Thus far, we have treated the contact resistances as linear, i.e. independent of the bias. This may be inappropriate for the actual contact resistances. In general the tunneling current increases strongly with the bias applied to a tunnel junction, and hence the contact resistances should therefore decrease significantly with increasing bias. A very simple model that accounts only for the population effects at energies relevant for injection (above the metal Fermi level by approximately Φ_B) leads to a bias dependence of the contact resistance of form, $R_{\uparrow,\downarrow}(\delta V) = R_{\uparrow,\downarrow}(0)\exp(-\delta V / kT)$. Here $\delta V \approx \Delta V$ is the voltage dropped across the contact barrier layer.[15] The actual bias dependence of the contact resistances is likely to be stronger. To explore the effect qualitatively, we assume $R_{\uparrow,\downarrow}(\delta V) = R_{\uparrow,\downarrow}(0)\exp(-\gamma\delta V / kT)$ and insert this type of dependence for the injecting and extracting contact resistances into the model and reexamine the spin polarization and MR. The resulting spin polarization and MR are plotted in figures 5 and 6. The values of the contact resistances are $R_{\uparrow}(0) = 2\times10^{-2}\Omega\text{cm}^2$ and $R_{\downarrow}(0) = 4\times10^{-2}\Omega\text{cm}^2$. It is evident, that current polarizations decrease relative to the case of constant contact resistances, but most striking result is the strong decrease of the MR with increasing bias. At high biases MR becomes negligibly small.

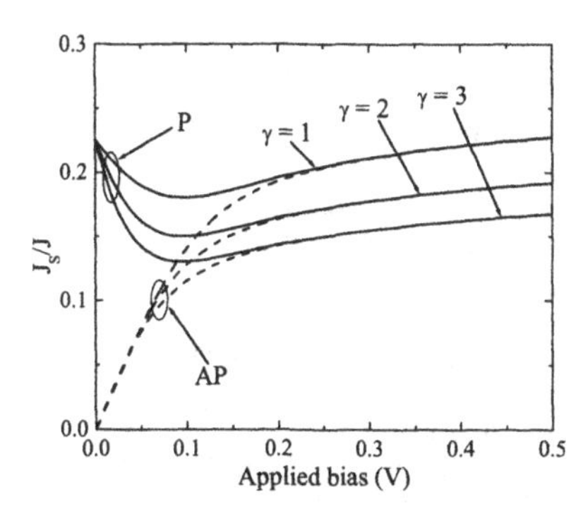

Figure 5. Spin polarization, J_S/J, for P and AP contact magnetization alignments. The contact resistances vary with bias as described in the text.

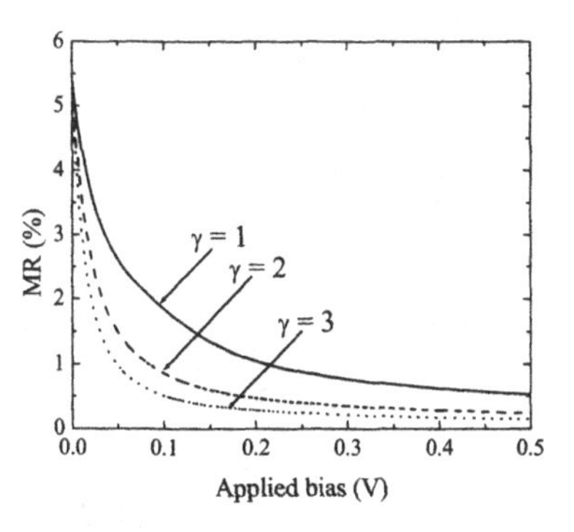

Figure 6. MR plotted for the different cases shown in figure 5.

CONCLUSIONS

Spin-selective injecting tunnel contacts can cause substantial spin polarization in organic semiconductors but measurable MR requires spin-selective tunnel contacts for both the injecting and the extracting electrodes. At low bias, both contacts control current polarization and MR may increase or decrease with the increasing applied bias depending on the contact resistances of the injecting and extracting contacts. At high bias, only the injecting contact controls spin injection. For linear contact resistances, spin polarization and MR increase with increasing applied bias. However, the increasing current densities with increasing applied bias imply increasing voltage drops across the contact tunnel barriers, which may result in their increased transparency or decreased contact resistance. This limits spin injection and strongly suppresses the MR with increasing bias.

ACKNOWLEDGMENTS

This work was supported in part by NSF (ECCS – 0724886). Work at Los Alamos National Laboratory was supported by DoE Office of Basic Energy Sciences Work Proposal No. 08SPCE973.

REFERENCES

[1] J. Shinar and R. Shinar, J. Phys. D: Appl. Phys. **41**, 133001 (2008).
[2] J. Fabian, A. Matos-Abiague, C. Ertler, P. Stano, and I. Žutić, Acta Phy. Slov., 57, pp. 565-907 (2007), and references therein.
[3] See for example: J. Daughton, J. Magn. Materials, **192**, 334 (1999)
[4] G. Schmidt, D. Ferrand, L. W. Molenkamp, A. T. Filip, and B. J. van Wees, Phys. Rev. B **62**, R4790 (2000).
[5] E. I. Rashba, Phys. Rev. B **62**, R16267 (2000).
[6] D. L. Smith and R. N. Silver, Phys. Rev. B **64**, 045323 (2001).
[7] P. P. Ruden and D. L. Smith, J. Appl. Phys. **95**, 4898 (2004).
[8] M. Yunus, P. P. Ruden, and D. L Smith, J. Appl. Phys **103**, 103714 (2008).
[9] B. T. Jonker, Proceedings of the IEEE, **91**, 727 (2003).
[10] S. Funaoka, I. Imae, N. Noma, and Y. Shirota, Synth. Met. **101**, 600 (1999), and T. Wangwijit, H. Sato, S. Tantayanon, Polym. Adv. Technol. **13**, 25 (2002).
[11] A. Fert, private communication.
[12] Z. H. Xiong, Di Wu, Z. Valy Vardeny, and J. Shi, Nature **427**, 821 (2004).
[13] Y. Liu, S. M. Watson, T. Lee, J. M. Gorham, H. E. Katz, J. A. Borchers, H. D. Fairbrother, and D. H. Reich, Phys Rev. B **79**, 075312 (2009).
[14] M. Yunus, P. P. Ruden, and D. L Smith, submitted to Synthetic Metals.
[15] S. O. Valenzuela, D. J. Monsma, C. M. Markus, V. Narayanamurti, M. Tinkham, Phys. Rev. Lett. 94, 196601 (2005), and D. L. Smith and P. P. Ruden, Phys. Rev. B 78, 125202 (2008).

Mater. Res. Soc. Symp. Proc. Vol. 1154 © 2009 Materials Research Society 1154-B10-26

Para-sexiphenyl-CdSe Nanocrystals Hybrid Light Emitting Diodes With Optimized Layer Thickness and Interfaces

C. Simbrunner[1], G. Hernandez-Sosa[1], E. Baumgartner[1], G. Hesser[2], J. Roither[1], W. Heiss[1] and H. Sitter[1]
[1] Institute of Semiconductor and Solid State Physics, Johannes Kepler University Linz, Austria
[2] Zentrum für Oberflächen und Nanoanalytik (ZONA), Johannes Kepler University Linz, Austria

ABSTRACT

CdSe/ZnS nanocrystals are embedded in para-sexiphenyl (p-6P) based hybrid light emitting diode devices providing red, green and blue (RGB) emission compatible to the HDTV color triangle. By structural and optical investigations the device parameters are optimized. The device performance is analyzed in respect to electrical and spectral response resulting in current-voltage characteristics with small leakage currents and low onset voltages. Furthermore the devices provide high color purity and stability which is demonstrated by their narrow emission line widths. All these results underline the ability of the presented device configuration to act as a future candidate for display applications.

INTRODUCTION

Para-sexiphenyl (p-6P) represents a well known candidate for the fabrication of high photon energy emitting devices [1,2]. The high energy gap (3.1 eV) of p-6P and consequently its blue electroluminescence (EL) emission recommends PSP as a component for multi-color organic light emitting device (OLED) displays and white LEDs. Recent publications promote solution-based semiconducting nanocrystals or quantum-dots (NC-QD) to be integrated within the OLED structure leading to a high efficient electroluminescence emission of organic-inorganic hybrid devices [3-4]. Such devices provide narrow emission bandwidth emerging from nearly monodisperse semiconductor quantum dots thus resulting in improvements with respect to color purity, color stability and color saturation. For commercial RGB multicolor display applications, a well standardized range of the visible spectrum has to be covered by the devices, as outlined by the corresponding coordinates in the chromaticity diagram according to the Commission Internationale de l'Eclairage (CIE). Whereas red and green emitting devices can be obtained by CdSe/ZnS NC-QDs, a proper blue emitting device is difficult to obtain with any kind of NCs [4,5]. Consequently p-6P based hybrid devices represent an interesting alternative providing the additionally needed blue electroluminescence by using the organic layer as actively emitting material in the absence of NC-QDs.

EXPERIMENTAL DETAILS

As indicated in fig. 1(a), all devices were built up on top of an indium tin oxide (ITO) coated glass substrate. To improve the hole injection efficiency of the ITO anode and to reduce the surface roughness, poly(3,4-ethylenedioxythiophene):poly(styrenesulfonate) (PEDOT:PSS) was spin coated onto the ITO and dried in vacuum for 24 h. For red (green) emitting devices, layers of CdSe/ZnS NC-QDs with CdSe core diameters of 4.8 (2.8 nm), respectively, were spin-coated from variously concentrated colloidal octane suspensions onto the anode. No NC-QDs were applied for the fabrication of blue emitting devices. The subsequent p-6P layer was prepared in a hot wall epitaxy (HWE) chamber maintained at a pressure of 9×10^{-6} mbar [6, 7]. In order to keep temperature gradients small during growth and to provide a clean and dry surface, the substrates were pre-heated at 100 °C for 15 min under vacuum conditions. After positioning the substrate above the hot wall oven for p-6P deposition, the p-6P source material was evaporated from a quartz tube heated to 240 °C, the wall temperature being 260 °C. For the optimized LED performances a growth time of 45 min was used. On top of this layer stack 100 nm of aluminum (Al) were deposited through a shadow mask as a cathode.

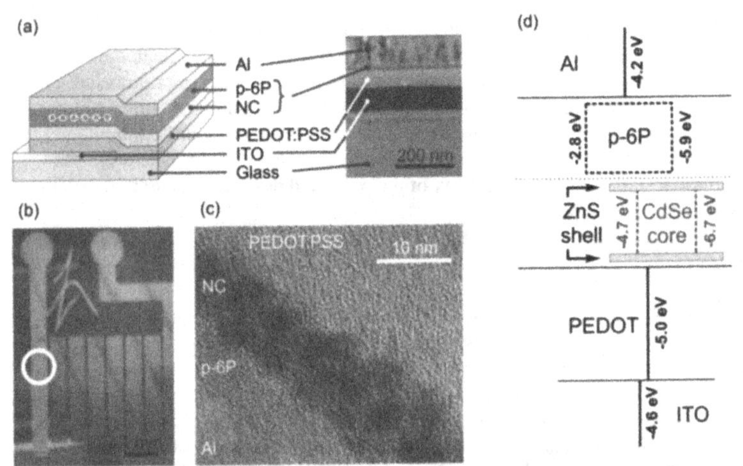

Figure 1. (a) Sketch of the hybrid LED sample structure and the corresponding cross sectional SEM image. (b) working device including NCs with a core diameter of 4.8 nm. (c) detailed TEM investigation of the active layer including 1-2 ML of NC in between PEDOT:PSS and p-6P layers (for red emitting devices with a NC-core diameter of 4.8 nm). (d) band level diagram for the hybrid LED structure used [8, 9].

The device homogeneity was analyzed by optical microscopy before and during operation as indicated in fig. 1(b) for a red emitting NC/p-6P hybrid device. All devices provide an active emitting rectangular area of 0.5x6 mm^2 and a line structure attached to it, representing the logo of our institute. As indicated in fig. 1(b), the tested devices exhibit uniform emission all over the active device in respect to intensity and color purity, underlining homogeneous current transport and carrier recombination. By using scanning electron microscopy (SEM) and transmission electron microscopy (TEM) the deposited layers were analyzed and optimized, as demonstrated by a typical SEM cross-section in fig. 1(a). In order to resolve the active emitting NC-QD/p-6P layers in more detail TEM provides a proper tool as demonstrated in fig. 1(c). It is found that the optimized layer thicknesses are less than 10 nm for the p-6P and for the NC-QD layer, corresponding to a coverage of 1-2 ML of NC-QDs. These observations are in very good agreement with hybrid devices based on other organic charge transport layers [4,10].

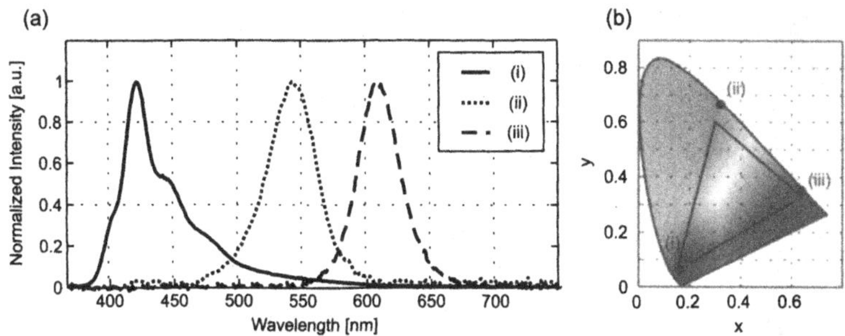

Figure 2. (a) Normalized electroluminescence spectra of p-6P only (i), green emitting NC (ii) and red emitting NC (iii) devices. (b) CIE chromaticity diagram showing the coordinates for (i) blue (0.16, 0.07), (ii) green (0.31, 0.66) and (iii) red emitting devices (0.64, 0.35).

By using a programmable I/V tracer in combination with a fiber-optical spectrometer the emitted electroluminescence spectra were investigated. Fig. 2(a) shows, normalized to their maximum intensity, the emission spectra, obtained at maximum driving current, of p-6P only based device (solid line), green emitting QD-NC (dotted line) and red emitting QD-NC (dashed line) devices. Strikingly, a high color purity is obtained for all three device types, quantified by a narrow emission band with a full width at half maximum (FWHM) of 40 nm for the QD-NC based hybrid devices and 20 nm for the devices solely acting with p-6P as active material. Furthermore in all spectra no parasitic emissions are observed and in particular the p-6P emission is completely quenched in the QD-NC devices. Consequently in these cases the p-6P solely acts as electron transport layer.

By calculating the corresponding CIE (1931) coordinates from the EL-spectra, the optical properties can be related to device applications. As indicated in fig. 2(b) the resulting CIE (1931) coordinates are given by (0.16, 0.07) for p-6P only (i), (0.31, 0.66) for green QD-NC (ii) and (0.64, 0.35) for red emitting QD-NC (iii) based devices. As indicated by the triangle in the CIE diagram these coordinates closely match with the standardized values for the HDTV display standard. Furthermore the high color purity of the devices is demonstrated in the diagram by their close vicinity to the monochromatic boundary.

In order to analyze the emission properties as a function of driving current in more detail, the obtained spectra are presented in fig. 3 as a contour plot showing the emission intensity as a function of driving current and emission wavelength. As indicated in fig. 3 (a) the p-6P device is dominated by a main emission peak located at 421 nm which continuously increases with rising device driving current. The corresponding I-V characteristic is depicted aside, providing a clear onset at about 4 V. An analogue behavior is observed for the NC-QD based devices as depicted in fig. 3 (b)-(c). For both NC-based devices the onset voltages are lowered with respect to the blue emitting p-6P device following the bandgap energies of the active emitting materials, indicating small losses.

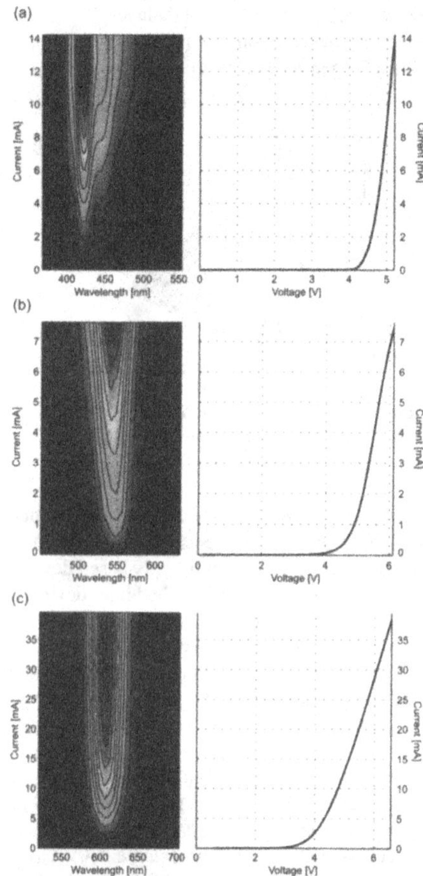

Figure 3. Contour plots of the electroluminescence spectra and applied voltage as a function of driving current for p-6P only (a), green emitting NC (b) and red emitting NC (c) devices.

CONCLUSIONS

In summary the successful integration of CdSe/ZnS core/shell NCs in p-6P based hybrid devices is presented. In particular the compatibility with the HDTV color triangle, which is covered by 94%, the high color purity and stability underline the presented device configuration as a future candidate for display applications. Furthermore the presented devices exhibit low onset voltages and negligible leakage currents due to the controlled deposition of compact, ultra thin organic p-6P layers by HWE.

ACKNOWLEDGMENTS

This work has been financially supported by the Austrian Fonds zur Foerderung der wissenschaftlichen Forschung FWF (projects NFN-S9706, Start Y179 and SFB 25-IRON).

REFERENCES

1. S. Tasch, C. Brandstätter, F. Meghdadi, G. Leising, G. Froyer, L. Althoel, Adv. Mat. 9 (1), 33 (1997)
2. G. Kranzelbinder, F. Meghdadi, S. Tasch, G. Leising, L. Fasoli, M. Sampietro, Syn. Metals 102, 1073 (1999)
3. S. Coe, W. Woo, M. Bawendi, V. Bulovic, Nature 420, 800 (2002)
4. Q. Sun, Y. Wang, L. Li, D. Wang, T. Zhu, J. Xu, C. Yang, Y. Li, Nature Photonics 1, 717 (2007)
5. A. Rizzo, Y. Li, S. Kudera, F. D. Sala, M. Zanella, W. J. Parak, R. Cingolani, L. Manna, and G. Gigli, Appl. Phys. Lett. 90, 051106 (2007)
6. A. Lopez-Otero, Thin Solid Films 3, 4 (1978).
7. G. Hernandez-Sosa, C. Simbrunner, T. Höfler, A. Moser, O. Werzer, B. Kunert, G. Trimmel, W. Kern, R. Resel and H. Sitter, Org. Electronics 10, 326 (2009)
8. G. Grem, V. Martin, F. Meghdadi, C. Paar, J. Stampfl, J. Sturm, S. Tasch, and G. Leising, Synth. Metals 71, 2193 (1995)
9. P. O. Anikeeva, C. F. Madigan, J. E. Halpert, M. G. Bawendi, and V. Bulovic, Phys. Rev. B 78, 085434 (2008)
10. A. L. Rogach, N. Gaponik, J. M. Lupton, C. Bertoni, D. E. Gallardo, S., Dunn, N. L. Pira, M. Paderi, P. Repetto, S. G. Romanov, C. O'Dwyer, C., M. S. Torres, and A. Eychmüller, Angew. Chem. 47, 6538 (2008).

Mater. Res. Soc. Symp. Proc. Vol. 1154 © 2009 Materials Research Society 1154-B10-38

Degradation of $Ir(ppy)_2(dtb\text{-}bpy)PF_6$ iTMC OLEDs

Velda Goldberg,[1] Michael D. Kaplan,[1] Leonard Soltzberg,[1] Dolly Armira,[1] Megan Bigelow,[1] Stephanie Bitzas,[1] Rachel Brady,[1] Shannon Browne,[1] Bianca Dichiaro,[1] Heather Foley,[1] Lauren Hutchinson,[1] Alison Inglis,[1] Nicole Kawamoto,[1] Amanda McLaughlin,[1] Caitlin Millett,[1] Hanah Nasri,[1] Sarah Newsky,[1] Tram Pham,[1] Cassandra Saikin,[1] Mary Scharpf,[1] Melissa Trieu,[1] George G. Malliaras,[2] and Stefan Bernhard[3]
[1]Chemistry and Physics Departments, Simmons College, Boston, Massachusetts; [2]Materials Science & Engineering Department, Cornell University, Ithaca, New York;
[3]Chemistry Department, Princeton University, Princeton, New Jersey

ABSTRACT

Simplicity of construction and operation are advantages of iTMC (ionic transition metal complex) OLEDs (organic light emitting diodes) compared with multi-layer OLED devices. Unfortunately, lifetimes do not compare favorably with the best multi-layer devices. We have previously shown for $Ru(bpy)_3(PF_6)_2$ based iTMC OLEDs that electrical drive produces emission-quenching dimers of the active species. We report evidence here that a chemical process may also be implicated in degradation of devices based on $Ir(ppy)_2(dtb\text{-}bpy)PF_6$ albeit by a very different mechanism. It appears that degradation of operating devices made with this Ir-based complex is related to current-induced heating of the organic layer, resulting in loss of the dtb-bpy ligand. (The dtb-bpy ligand is labile compared with the cyclometallated ppy ligands.) Morphological changes observed in electrically driven $Ir(ppy)_2(dtb\text{-}bpy)PF_6$ OLEDs provide evidence of substantial heating during device operation. Evidence from UV-vis spectra in the presence of an electric field as well as MALDI-TOF mass spectra of the OLED materials before and after electrical drive add support for this model of the degradation process.

INTRODUCTION

As OLED-based displays begin to appear in major mainstream consumer devices, such as television sets, the importance of understanding the mechanisms of OLED degradation increases. Chemical processes, such as decomposition or other reactions of the active compounds, and physical processes, such as delamination of metal contacts have been implicated in various studies. In either case, electrical drive is likely to cause, or at least accelerate, device degradation.

Iridium-based iTMC compounds are among the brightest emitters in the OLED palette. However, like all other known OLED materials, these compounds are also subject to performance degradation during operation. We have accumulated increasing evidence that degradation in the bright yellow emitter $Ir(ppy)_2(dtb\text{-}bpy)PF_6$ (Figure 1) results from chemical decomposition caused by substantial heating during electrical drive [ppy = 2-phenylpyridyl and dtb-bpy = 4,4'-di-*tert*-butyl-2,2'-dipyridyl].

Our approach to these studies has been to correlate electrical and light emission measurements with MALDI-TOF mass spectra of the OLED material before and after device operation. These observations have been supplemented, and the suggested degradation mechanism to some extent corroborated, by UV-vis spectra recorded in the presence of an electric field.

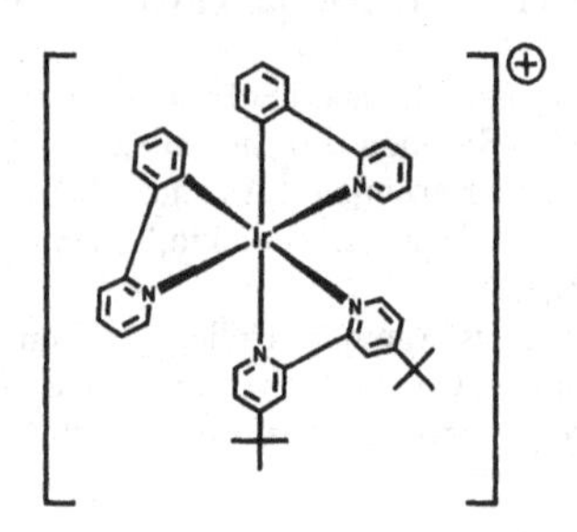

Figure 1. $Ir(ppy)_2(dtb\text{-}bpy)^+$ cation.

EXPERIMENTAL DETAILS

Devices consist of the organic material sandwiched between an ITO (indium tin oxide) anode and an aluminum, silver or gold cathode. All electrical measurements are conducted under ambient conditions, and our methods for preparing, operating, and monitoring OLEDs have been reported previously [1]. In contrast with the widely-studied ruthenium complex $Ru(bpy)_3(PF_6)_2$, which degrades via formation of a dimeric complex [2], mass spectra from OLED devices fabricated with the iridium complex $Ir(ppy)_2(dtb\text{-}bpy)PF_6$ [3] and run to extinction show no evidence of species at higher mass than the parent complex. Instead, positive ion mass spectra of material from devices subjected to electrical drive show a decrease in abundance of the emissive cation at m/z 767/769, and negative ion mass spectra indicate the disappearance of the counterion, PF_6^-, at m/z 145. In such devices, a new group of peaks appears in the range m/z 550-700, spaced by 22 mass units. These species do not contain iridium, because they do not show the Ir isotope pattern.

A positive ion MALDI-TOF spectrum in the range m/z 550-700 for the newly synthesized OLED material is shown in Figure 2. The parent complex is clearly visible at m/z 767/769; the inset shows the characteristic isotope pattern for this complex. Figure 3 shows the spectrum of material taken from device #1 (Al cathode) of another sample after running for 24 hours at 7.0 Volts. (The inset in Figure 3 indicates the device configuration used for all samples.) The appearance of the six prominent peaks in the range m/z 550-700 is indicative of decomposition of the organic material, and this may partially be due to excess heat generated during device operation. All samples for MALDI-TOF mass spectra were prepared with the matrix 6-aza-2-thiothymine (ATT), $C_4H_5N_3OS$, with a nominal mass of 143 Da. The peak at m/z 569 is consistent with the composition $[dtb\text{-}bpy + PF_3 + PF_2 + ATT + H]^+$. This species contains the non-cyclometallated ligand, fragments of the PF_6^- counterion, and a molecule of the matrix. The masses of the succeeding peaks at m/z 591, 613, 635, 657, 679 correspond to successive adduction of sodium ions. (Matrix adducts are common in MALDI-TOF mass spectrometry [4], as is cation formation via protonation. Sodium adducts are also common in MALDI-TOF spectra because of the ubiquitous presence of sodium in most matrices. In all cases, observed species are singly-charged [5].)

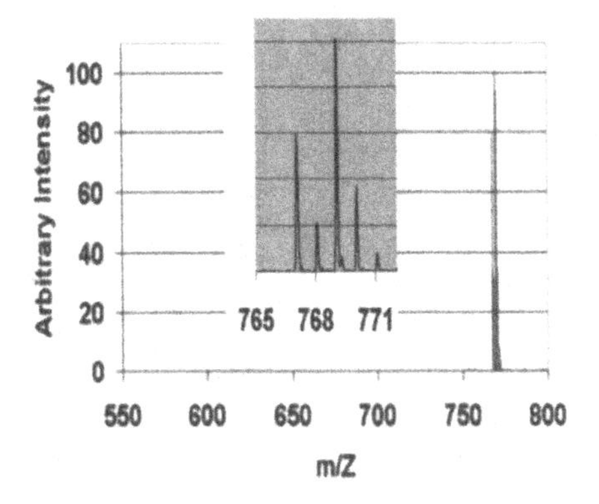

Figure 2. Normalized MALDI-TOF spectrum showing positive ion peaks of the newly synthesized starting material $Ir(ppy)_2(dtb\text{-}bpy)PF_6$. Inset shows characteristic isotope pattern due to iridium.

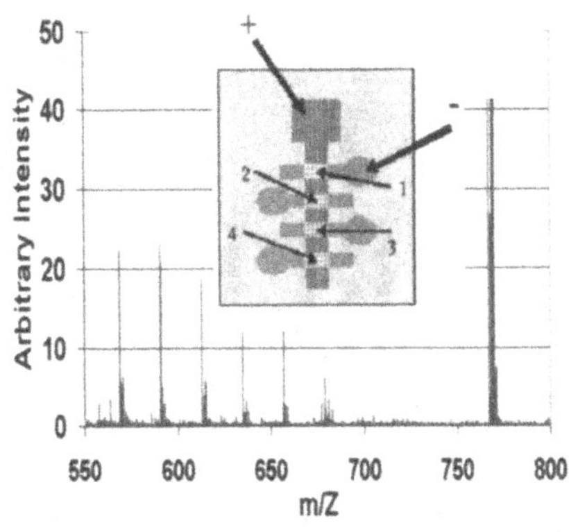

Figure 3. Normalized MALDI-TOF spectrum for device #1 (Al cathode) after running for 24 hours at 7.0 Volts. The signal at m/z 767/769 is due to the parent compound. The inset shows the configuration for devices #1-4 on a typical sample (scale: 25 mm x 25 mm).

These results and those of several other experiments that suggest the role of excessive heating are summarized in Figures 4–6. Figure 4 compares the relative, normalized maximum intensities for each of the six peaks for device #1 (based on the spectrum shown in Figure 3) with those for devices #2 and #4 on the same slide. In this case, only device #1 was run, so the peaks seen in spectra for devices #2 and #4 are presumably due to conductive heat transfer along the ITO strip. Note that the peaks are least pronounced in device #4, which is situated furthest from the operating device.

We additionally tested the role of conductive heat transfer by running all four devices simultaneously at 7.0 Volts for 24 hours. Due to the device configuration (as shown in the overlay in Figure 3) and the branching of the current through each device in turn, device #1 should be subject to the greatest heating and #4 to the least. Indeed, as shown in Figure 5, the peaks are most pronounced in device #1 and least pronounced in device #4. The parent complex (m/z 767/769) is also substantially diminished in device #1. (This set of devices had Au cathodes.)

To test the effect of proximity heating in the absence of electric drive, the top electrode of device #1 on a separate sample was heated with a hot tip for 24 hours. Separate spectra were then recorded for material taken from this region and from region #2. Figure 6 compares the relative peak intensities for these devices, and, once again, there is a pronounced proximity effect.

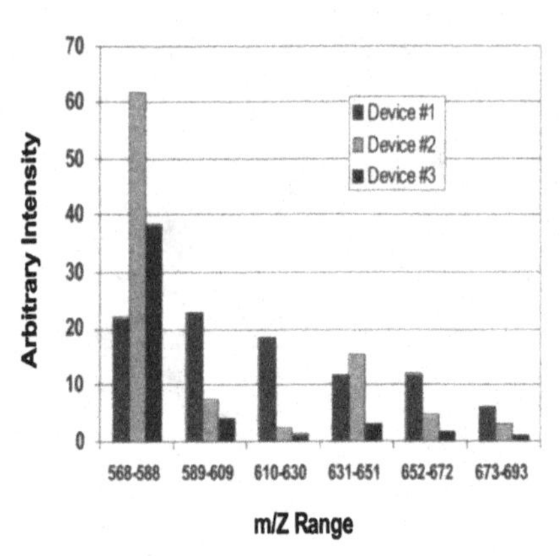

Figure 4. Normalized MALDI-TOF maximum intensity comparison for a series of peaks for run device #1, adjacent device #2, and twice removed device #4—all having Al cathodes.

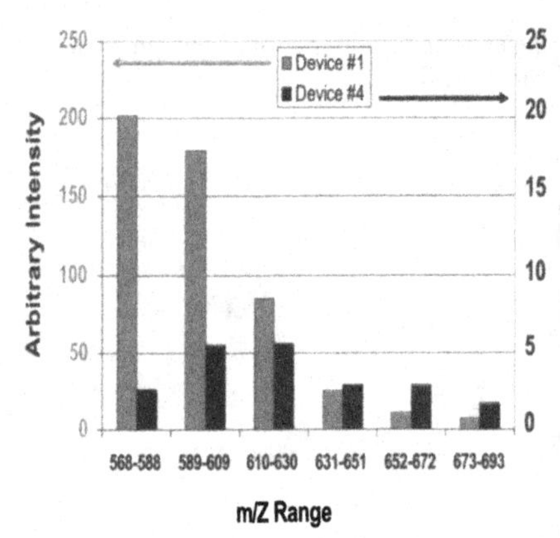

Figure 5. Normalized MALDI-TOF maximum intensity comparison for a series of peaks for device #1, subjected to the greatest heating, and device #4, subjected to the least heating, when all four devices having Au cathodes are run in parallel.

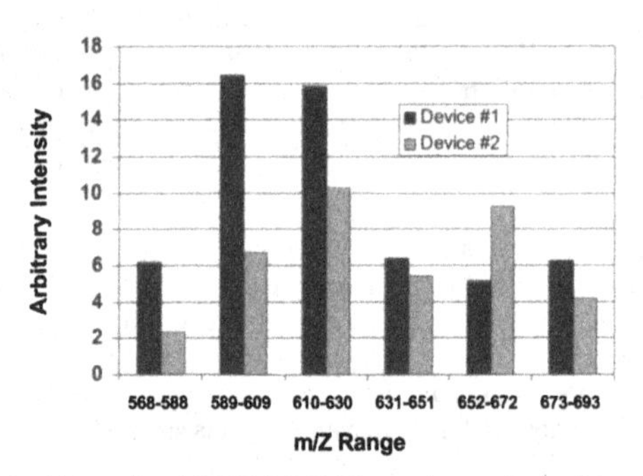

Figure 6. Normalized MALDI-TOF maximum intensity comparison for a series of peaks for device #1 heated, and adjacent device #2, both with Al cathodes.

To investigate a possible catalytic role of the metal cathode in the decomposition of the OLED material, MALDI-TOF mass spectra of the iridium complex mixed with certain powdered metals were recorded. These spectra show loss of the di-*tert*-butyl bipyridyl ligand from the complex (Figure 7B). The addition of aluminum nitrate to both samples shows that the effect is due to the zero-valent metal and not simply a generic presence of aluminum. This effect is especially pronounced with aluminum and is also observed with copper, tin, and some other metals, though not with silver or titanium. In these experiments, energy from the MALDI laser rather than heating or electrical drive is the likely energy source for the decomposition.

The electric field may also play a role during device operation. The UV spectra in Figure 8 from the Ir-complex dissolved in acetonitrile demonstrate sensitivity of the OLED compound to an external field, although it is not clear whether this change is connected with device degradation. The peak at 257 nm and its shoulder at longer wavelength are associated with absorption by the ligands. The electric field influence could be related to several factors, such as electrochemical processes in the solution, change of the electric dipole moment of the complex, or heat induced changes of the materials; the observed spectra may be consistent with loss of the dtb-bpy ligand from the complex. More data will be necessary to refine this picture.

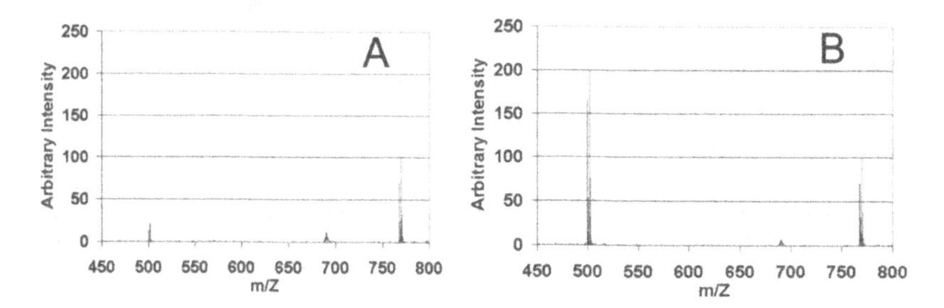

Figure 7. (A) MALDI-TOF spectrum of $Ir(ppy)_2(dtb\text{-}bpy)PF_6$ mixed with $Al(NO_3)_3$; small $Ir(ppy)_2^+$ peak at m/z 499/501 indicates modest loss of dtb-bpy from complex. **(B)** Addition of Al^0 powder greatly increases loss of dtb-bpy.

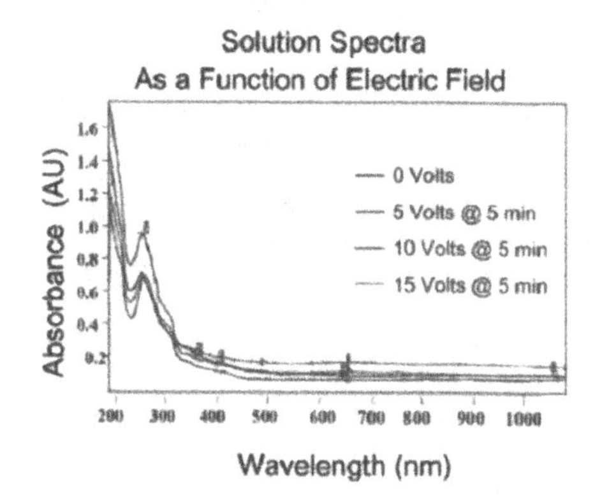

Figure 8. UV-vis spectra of the Ir-complex dissolved in acetonitrile and subjected to increasing electric field strength.

DISCUSSION AND CONCLUSIONS

It seems clear that the degradation mechanism for iridium-based iTMC OLEDs differs from that of ruthenium-based devices. While dimerization produces effective quenching centers in the ruthenium-based material, it appears that chemical decomposition is at work in the iridium devices. Both the emissive cation and the mobile anion responsible for establishing the space-charge needed for charge injection disappear from devices subjected to electrical drive or to heat. Destruction of the emissive cationic complex as well as the mobile anion would, of course, ruin the device. It seems plausible, then, that heat generated in operating devices leads to decomposition of the active material, possibly catalyzed by the cathode metal. Experiments are underway to test this hypothesis further.

In terms of device performance, the heating implications suggest that the choice of substrate and overall sample architecture may ultimately play an important role in mitigating degradation processes due to heating effects.

ACKNOWLEDGMENTS

We acknowledge the support of the National Science Foundation (CHE-0216268 and DMR-0605621). The Cornell work was supported by the National Science Foundation (DMR0094047), the Cornell Center for Materials Research (CCMR), and NYSTAR.

REFERENCES

1. V. Goldberg, M.D. Kaplan, L.J. Soltzberg, H. Bankowski, S. Browne, H. Concannon, M. Damour, S. Green, E. Hendrickson, H. Huang, V. Liu, L. Piirainen, S. Reel, G.G. Malliaras, J.D. Slinker, and S. Bernhard: Degradation in iTMC OLEDs, in *Interfaces in Organic and Molecular Electronics III* edited by Karen L. Kavanagh (*Mater. Res. Soc. Symp. Proc.* **1029E**, Warrendale, PA, 2007), F3.30.
2. L.J. Soltzberg, J.D. Slinker, S. Flores-Torres, D.A. Bernards, G.G. Malliaras, H.D. Abruña, J-S. Kim, R.H. Friend, M.D. Kaplan and V. Goldberg, *J. Amer. Chem. Soc.* **128**, 7761–7764 (2006).
3. J.D. Slinker, A.A. Gorodetsky, M.S. Lowry, J. Wang, S. Parker, R. Rohl, S. Bernhard, and G.G. Malliaras, *J. Amer. Chem. Soc.* **126**, 2763-2767 (2004).
4. L.J. Soltzberg, K. Do, S. Lokhande, S. Ochoa, and M. Tran, *Rapid Communications in Mass Spectrometry* **19**, 2473-2479 (2005).
5. M. Karas, M. Glückmann and J. Schafer, *Journal of Mass Spectrom*etry **35**, 1–12 (2000).

Mater. Res. Soc. Symp. Proc. Vol. 1154 © 2009 Materials Research Society 1154-B10-40

Comparison of Molecular Monolayer Interface Treatments in Organic-Inorganic Photovoltaic Devices

Jamie M. Albin[1], Darick J. Baker[1], Cary G. Allen[1], Thomas E. Furtak[1], Reuben T. Collins[1,2], Dana C. Olson[3], David S. Ginley[3], Christian C. Weigand[4], Astrid-Sofie Vardoy[4], Cecile Ladam[5]
[1]Colorado School of Mines, Golden, CO; [2]Renewable Energy Materials Research Science and Engineering Center, Golden, CO; [3]National Renewable Energy Laboratory Golden, CO; [4]Norwegian University of Science and Technology, Trondheim, Norway; [5]SINTEF, Trondheim, Norway

ABSTRACT

In this study, we explore the effects of alkyl surface terminations on ZnO for inverted, planar ZnO/poly(3-hexylthiophene) (P3HT) solar cells using two different attachment chemistries. Octadecylthiol (ODT) and octadecyltriethoxysilane (OTES) molecules were used to create 18-carbon alkyl surface molecular layers on sol gel-derived ZnO surfaces. Molecular layer formation was confirmed and characterized using water contact angle measurements, infrared (IR) transmission measurements, and X-ray photoelectron spectroscopy (XPS). The performances of the ZnO/P3HT photovoltaic cells made from ODT- and OTES-functionalized ZnO were compared. The ODT-modified devices had higher efficiencies than OTES-modified devices, suggesting that differences in the attachment scheme affect the efficiency of charge transfer through the molecular layers at the treated ZnO surface.

INTRODUCTION

Excitonic hybrid organic-inorganic solar cells are gaining viability as alternatives to p-n junction photovoltaics. Although hybrid cells typically have lower efficiencies than their inorganic counterparts, they are more compatible with inexpensive manufacturing techniques such as spray deposition and roll-to-roll processing, which can reduce the fabrication cost per photovoltaic watt. Polymer devices with nanostructured ZnO as the electron-accepting layer have the potential to improve carrier collection and power conversion efficiency in the bulk heterojunction approach to organic solar cells [1]. The ZnO/polymer interface, however, is not optimal and properties such as polymer ordering, wetting at the interface, and charge transfer across the interface need improvement. Functionalization of the ZnO surface with molecular monolayers has the potential to resolve these issues [2,3]. Monson (Hsu is not the first author) et al. observed increased current density with alkyl chain length in alkylthiol-modified ZnO/P3HT devices despite the fact that the molecules are expected to lengthen the electron tunneling barrier [2].

In this study, we demonstrate formation of surface molecular layers on sol gel-derived ZnO using octadecylthiol (ODT) and octadecyltriethoxysilane (OTES). Both molecules create an 18-carbon chain alkyl surface termination allowing a comparison of ZnO surfaces functionalized with the same end group but using different attachment chemistries. Molecular layer formation was characterized using water contact angle measurements, infrared (IR) transmission measurements, and X-ray photoelectron spectroscopy (XPS). The performance of inverted planar ZnO/poly(3-hexylthiophene) (P3HT) photovoltaic cells made from ZnO films functionalized using ODT and OTES were compared. Visible absorption measurements indicated that the P3HT layers for both treatments had improved order relative to P3HT spun onto untreated ZnO. The ODT-modified devices, however, had higher efficiencies than OTES-treated devices, which showed decreased short circuit current. These observations suggest that differences in the attachment chemistry affect the efficiency of charge transfer through the molecular layers at the treated ZnO surface.

EXPERIMENT

The substrates used were 1 in^2 sections of silicon (for transmission infrared (IR) spectroscopy), glass slides (for UV-visible spectroscopy), and glass slides with patterned indium tin oxide (ITO) (used in photovoltaic devices). The substrates were cleaned in ultrasonic baths of acetone then isopropyl alcohol for 15 min. each, dried with nitrogen, then oxygen plasma treated at 155 W for 5 min. The ZnO sol gel solution was created by stirring 0.2 mL ethanolamine, 820 mg zinc acetate, and 5 mL 2-methoxyethanol at 60 C for 30 min until the acetate was dissolved. The solution was spun-cast onto prepared substrates (2000 rpm, 50 sec) and annealed at 300 C for 10 min in air. The films were cooled to room temperature, rinsed with deionized water, acetone, and isopropyl alcohol, and then blown dry with nitrogen. Typical ZnO films have RMS roughness of approximately 2 nm. The films were then functionalized with either ODT or OTES, or received no molecular treatment. Immediately prior to molecular treatments, the substrates were annealed at 150 C for 30 min in air to drive off water and cooled to room temperature

For the OTES treatment, the ZnO films were hydroxylated in a UV-ozone cleaner (Jelight, CA) for 15 minutes, then soaked in a "fresh" 1 mM solution of octadecyltriethoxysilane (Gelest, PA) in toluene with 0.5 mL n-butylamine kept at 45 C for 90 minutes, according to the procedure in Refs [4] and [5]. The samples were then removed, rinsed with toluene and acetone and blown dry with nitrogen. The samples were then annealed at 100 C in a vacuum oven for 60 minutes, rinsed with acetone and blown dry.

For the ODT treatment, the annealed ZnO films were soaked in a 1 mM solution of ODT in ethanol for 72 hours, according to the procedure used by Monson et al. [2]. The samples were then removed, rinsed with ethanol and blown dry with nitrogen.

Water contact angles (CA) were determined from images of sessile water drops taken from several points on these films and analyzed with polynomial spline image analysis software [6]. IR transmission spectra were obtained with a Nicolet Magna Jr. 560 FTIR using a calcium fluoride beam splitter and a liquid N_2-cooled HgCdTe detector. XPS measurements were obtained using a Kratos Analytical (Manchester, U.K.) HSi system and a monochromated Al Kα beam source. Power conversion efficiencies of the devices were measured in air under AM1.5 illumination using a Spectrolab XT-10 solar simulator.

UV-Visible spectroscopy was performed on ~10 nm thick P3HT (Reike) films deposited onto the ZnO layers from a 1 g/L chloroform solution. The solution was spun-cast in air at 2000 rpm for 60 seconds onto freshly functionalized ZnO films, or freshly annealed films in the case of untreated samples.

Photovoltaic cells were fabricated by spin-casting P3HT films at 800 rpm for 60 sec in air from 15 g/L chloroform solution onto ZnO films prepared on indium tin oxide (ITO) on glass.. This deposition produced P3HT films ~80 nm thick. The samples were then transferred to an evaporator in a nitrogen glove box where a 100 nm Ag back-electrode was evaporated at 2 Å/sec onto the P3HT layer through a patterned shadow mask. The resulting active device area was 0.11 cm^2. The structure of these devices is illustrated in figure 1.

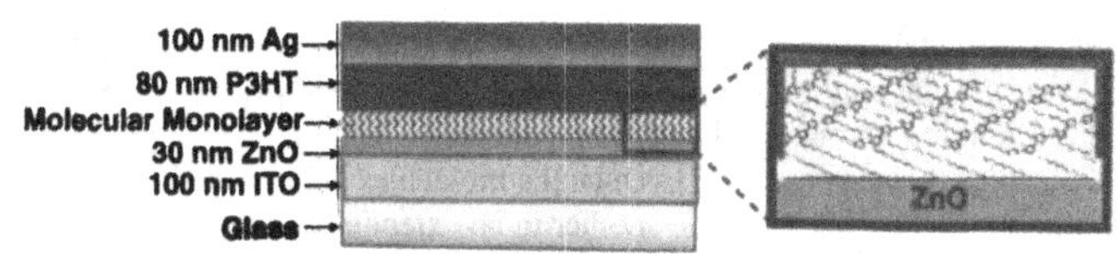

Figure 1. ZnO/18-carbon/P3HT photovoltaic device structure.

RESULTS and DISCUSSION

Monolayer Characterization

Typical measured water CA values were 120° for ODT-treated ZnO and 107° for OTES-treated ZnO. While the absolute contact angles can be influenced by the roughness of the underlying ZnO surface, our results for the sol gel films with molecule-treated surfaces as described here are reproducible with error bars of a few degrees and hence useful for comparison of various treatments. Untreated ZnO films yield less reproducible CA values depending upon handling history. The films are usually hydrophilic and yield contact angles of approximately 65° with large error bars. The dramatic increase in CA after molecular treatment can be attributed to the reduction of surface energy by the hydrophobic alkyl surface terminal groups of the functional molecules.

The ODT-treated samples may yield higher CA values than OTES-treated samples for several reasons. The ODT may have higher surface coverage, exposing a smaller area of the polar ZnO surface. The ODT molecules could have higher molecular tilt that the OTES molecules, which would expose a larger cross-section of non-polar methylene groups in the alkyl chains. The steps in the attachment process may also play a role. For example, polar hydroxyl groups that remained on the surface after OTES deposition could reduce contact angle. Differences in surface coverage may be better understood by examining the differences in the ways that triethoxysilane and thiol attachment groups bind to surfaces. The proposed deposition mechanisms are shown in figure 2. We believe that alkylthiols self-assemble on Zn sites much the same way that they do on gold, resulting in a final structure of close-packed alkyl chains aligned at a small angle with respect to the surface normal [8]. In the case of the OTES

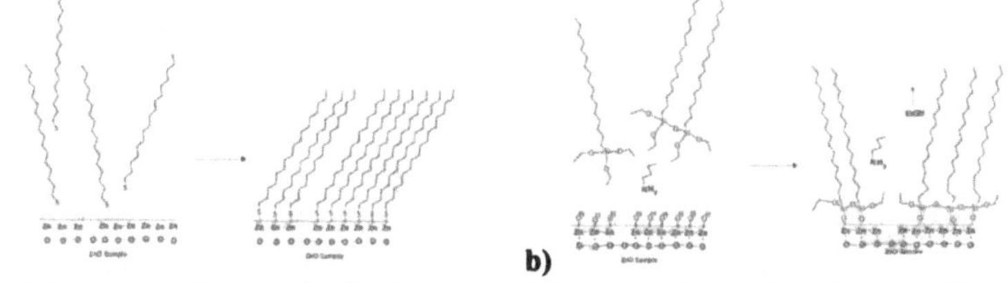

Figure 2. Diagrams of how molecular layers might form on a Zn-terminated ZnO surface. (a) ODT: During solution deposition, molecules diffuse to the ZnO surface, then mobilize into the lowest energy configuration of densely packed alkyl chains. (b) OTES: Molecules in solution

diffuse to the ZnO surface. The final monolayer may be fully or partially cross-linked through siloxane bonds, creating a 2-D network of molecules across the surface.

deposition, the n-butylamine-catalyzed surface reaction binds an attachment group from the OTES molecule at a hydroxyl site on the ZnO surface through a Si-O bond [9]. However, the OTES molecules can cross-link with one another to some degree in solution, which produces inhomogeneities in molecular surface density [7]. Once the molecules are attached to the surface, they may continue to cross-link with one another in an extended 2D network of siloxane bonds. XPS spectra of OTES-treated films exhibit a characteristic Si 2p peak at 103 eV. This is evidence of siloxane bonding between the molecule and the surface or between cross-linked OTES molecules.

Infrared absorbance spectra for ODT and OTES treated ZnO layers are shown in figure 3a. Peak locations give indications of the alkyl chain ordering, where lower energies (lower wavenumbers) correspond to higher ordering. Reported values of well-ordered C-H symmetric modes (ν_s) are from 2846 to 2850 cm^{-1}, and values for asymmetric modes (ν_a) are from 2915 to 2920 cm^{-1}. The ODT peaks are $\nu_s = 2850\ cm^{-1}$ and $\nu_a = 2919\ cm^{-1}$, indicative of high alignment among the alkyl chains. The OTES peaks are shifted to higher energies of $\nu_s = 2855\ cm^{-1}$ and $\nu_a = 2925\ cm^{-1}$, well outside the ranges for ordered layers. The OTES peaks are between those of an ordered layer and liquid-like (disordered) monolayer [4]. Gauche defects as well as inhomogeneities in surface coverage from incomplete cross-polymerization between OTES molecules may contribute to the observed conformational disorder of the alkyl chains [4,7]. The greater integrated intensity of the ODT peaks also suggests a more dense layer. However, interpretation of the integrated intensities to obtain coverage is subtle and still under investigation.

P3HT Ordering

In a ZnO/P3HT organic solar cell, highly ordered P3HT is desirable for optimal photocurrent collection [10]. In this study, ordering is qualitatively characterized through the presence and definition of sharp peaks and shoulders in UV-visible absorption spectra, attributed to intrachain and interchain π-π orbital coupling in ordered domains. Shifts in the spectral features toward red wavelengths are also indicative of ordering. UV-visible absorption spectra shown in figure 3b indicate that both molecular treatments resulted in more ordered films as compared to films deposited onto untreated ZnO.

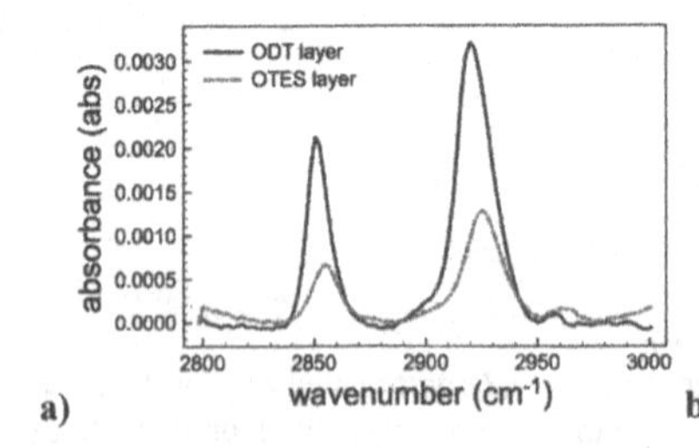

a)

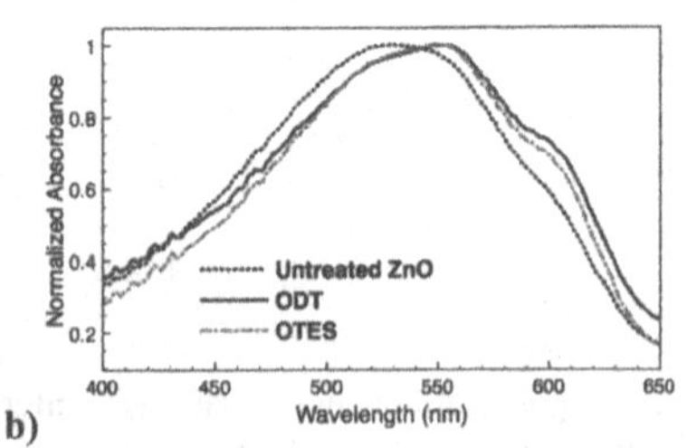

b)

Figure 3. (a) IR reflection absorbance spectra of the asymmetric and symmetric C-H stretch modes of the SAMs. (b) UV-Vis spectra for comparison of ordering in 10 nm P3HT films spun onto untreated ZnO and ZnO treated with ODT or OTES.

Devices

Device results are presented in figure 4 and Table I. It is noteworthy that although both molecular treatments enhanced polymer ordering, the OTES treatment reproducibly resulted in lower V_{oc} and J_{sc} than the ODT treatment. Although in this set of devices the ODT treatment gave enhanced Jsc relative to the control, within experimental error, the ODT treatment and control prepared as described above gave comparable results. The performance characteristics of our ODT-treated and untreated devices show similar trends to reports in the literature [2]. However, the poor performance of the OTES-treated device indicates that P3HT ordering at the interface is just one of many considerations in ZnO/P3HT device performance.

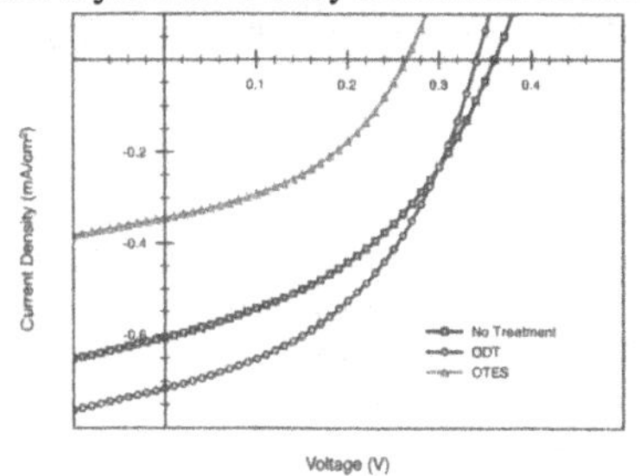

Figure 4. I-V curves of treated and untreated samples under illumination.

Table I: Current-voltage characteristics of devices made with and without SAM modification.

Surface Treatment	Voc (V)	Jsc (mA/cm²)	FF	η
No Treatment	0.361	0.6062	41.4%	0.09%
ODT	0.341	0.716	43.9%	0.11%
OTES	0.264	0.3453	41.7%	0.04%

Among the possible explanations for differences in performance between ODT and OTES-treated devices are the different attachment species, orientation/ordering of the alkyl chains, and the differences in surface preparation before the molecular deposition. For example, the S-Zn bond formed with the ODT treatment may be more conducive to charge transfer than the siloxane bond because of the bonding characteristics and/or the interface energetics arising from differences in the overall surface dipole created with the different attachment groups. There may also be an effect from the structure of the alkyl chains if transport is affected by the more ordered ODT vs. disordered OTES alkyl chains.

It should also be noted that both the ODT and OTES treatments gave more reproducible device characteristics from one preparation to the next than the control device preparation process. This suggests that molecular treatments reduce variability in surface properties, such as defect state density, that affect device performance.

CONCLUSIONS

Both the ODT and OTES molecular treatments resulted in alkyl terminated ZnO surfaces that reduced the surface energy of ZnO films and improved ordering of the P3HT chains at the organic/inorganic interface, but neither treatment significantly improved device performance characteristics. Compared to untreated devices, the ODT treatment produced devices comparable to the control, and the OTES treatment reduced V_{oc} and J_{sc}. This leads us to believe that the differences in device performance were more attributable to the effect of the molecular attachment scheme on ZnO surface properties (processing, SAM bonding species, alkyl chain ordering, etc.), rather than the morphology of the P3HT. Regardless of performance, the improved reproducibility of both types of functionalized devices demonstrates the significant effects of functional molecules on the surface properties of ZnO.

ACKNOWLEDGMENTS

The authors would like to thank Joe Berry, Tom Reilly, Matt Reese, Matt White, Matt Bergren and Brian Gregg for their help and advice. This work was supported by the National Science Foundation under grant no. DMR-0606054 and DMR-0820518.

REFERENCES

1. W. J. E. Beek, M. M. Wienk, M. Kemerink, X.Yang and R. A. J. Janssen, *J. Phys. Chem. B* **109**, 9505-9516 (2005).
2. T. C. Monson, M. T. Lloyd, D. C. Olson, Y. Lee, and J. W. P. Hsu, *Adv. Mater.* **20**, 4755 (2008).
3. C. Goh, S. R. Scully, and M. D. McGehee, *J. Appl. Phys.* **101**, 114503 (2007).
4. C. G. Allen, D. J. Baker, J. M. Albin, H. E. Oertli, D. T. Gillaspie, D. C. Olson, T. E. Furtak, and R. T. Collins, *Langmuir* **24,** 13393-13398 (2008).
5. D. M. Walba, C. A. Liberko, E. Korblova, and M. Farrow, *Liq. Crys.* **31,** 481-489 (2004).
6. A. F. Stalder, G. Kulik, D. Sage, and P. Hoffmann, *Colloids Surf. A* **286,** 92-103 (2006).
7. S. Onclin, B. J. Ravoo, and D. N. Reinhoudt, *Angew. Chem. Int. Ed.* **44,** 6282-6304 (2005).
8. D. M. Rosu, J. C. Jones, J. W. P. Hsu, K. L. Kavanagh, D. Tsankov, U. Schade, N. Esser, and K. Hinrichs, *Langmuir* **25,** 919-923 (2009).
9. L. D. White and C. P. Tripp, *J. Colloid Interface Sci.* **227,** 237-243 (2000).
10. D. C. Olson, Y. Lee, M. S. White, N. Kopidakis, S. E. Shaheen, D. S. Ginley, J. A. Voigt, and J. W. P. Hsu, *J. Phys. Chem. C.* **111,** 16640-16645 (2007).

Mater. Res. Soc. Symp. Proc. Vol. 1154 © 2009 Materials Research Society 1154-B10-45

Optical Stability of Small-Molecule Thin-Films Determined by Photothermal Deflection Spectroscopy

M. Stella[1], M. Della Pirriera[2], J. Puigdollers[2], J. Bertomeu[1], C. Voz[2], J. Andreu[1], R. Alcubilla[2].
1. Dep. Física Aplicada i Òptica. Universitat de Barcelona
2. Micro and Nanotechnology Group, Dept Enginyeria Electrònica, Universitat Politècnica Catalunya

ABSTRACT

In this paper the optical absorption properties of n-type C_{60} and PTCDA, and p-type CuPc small molecule semiconductors are investigated by optical transmission and Photothermal Deflection Spectroscopy (PDS). The results show the usual absorption bands related to HOMO-LUMO transitions in the high absorption region of the transmission spectra. PDS measurements also evidences exponential absorption shoulders with different characteristic energies. In addition, broad bands in the low absorption level are observed for C_{60} and PTCDA thin-films. These bands have been attributed to contamination due to air exposure. In order to get deeper understanding of the degradation mechanisms single and co-evaporated thin-films have been characterized by PDS. The dependence of the optical coefficient on exposure to light and air have been studied and correlated to the structural properties of the films (as measured by X-Ray Diffraction Spectroscopy). The results show that CuPc and PTCDA are quite stable against light and air exposure, while C_{60} shows important changes in its absorption coefficient. The bulk heterojunctions show stability in agreement with what observed for single layers, since the absorption coefficient of CuPc:PTCDA is almost not altered after the degradation treatments, while CuPc:C_{60} shows changes for low energy values.

INTRODUCTION

Organic semiconductors represent a new interesting class of materials for several electronic applications. Organic solar cells performance have improved significantly in the last few years thanks to the optimization of the solar cell structure and, specially, to the ability to process new organic semiconductors with optimised properties.

The conventional transmission and reflection spectroscopies are not suitable for detecting low level absorptions because their sensibility is in the order of $\alpha d = 10^{-2}$, so Photothermal Deflection Spectroscopy (PDS) has been used in this work. PDS is a technique known for its highly sensitive and non-destructive capability of determining the optical absorption of solids, liquids or gases by employing the photothermal effect.

P-type semiconductor materials, as like as copper phtalocyanine (CuPc) and *n*-type ones, as like as fullerene (C_{60}) and perylene-3, 4, 9, 10-tetracarboxylic-3,4,9,10-dianhydride (PTCDA), are organic semiconductors that are commonly used for organic electronic applications [1-4]. The combination of these materials in coevaporated thin films is used for photovoltaic studies [5, 6]. PDS technique was used to know the optical properties of these materials [7] and X-Ray Diffraction spectroscopy (XRD) was used to analyze the amorphous and crystalline phase in the films [8]. In order to know the degradation under illumination and air exposure conditions, some previous characterization have been realized studying their optical and structural properties by PDS and XRD respectively [9, 10].

The optical gap and Urbach energy can be calculated from the absorption coefficient using adequate models, depending on the kind of sample. In the case of polycrystalline samples, the optical gap is fitted from the standard model used for direct allowed transitions in inorganic semiconductors. The equation employed is the following [11]:

$$\alpha(h\upsilon) = B \cdot (h\upsilon - E_{og})^{1/2} \quad (1)$$

where B is a constant and E_{og} the optical gap. Such model comes from the general theory on inorganic semiconductors and only a few examples can be found in literature about its employment with organic ones [12]. In the case of amorphous thin-films the Tauc equation will be evaluated and its result compared with the one obtained by equation (1). Tauc's equation can be written, in the case of parabolic band edges, in the following way [13]:

$$[\alpha(h\upsilon) \cdot h\upsilon]^{1/2} = A \cdot (h\upsilon - E_{og}) \quad (2)$$

where A is a constant. The Urbach energy can be calculated as follows [14]:

$$\alpha(h\upsilon) = \alpha_0 \exp\left(\frac{h\upsilon}{E_u}\right) \quad (3)$$

where α_0 is a constant, $h\upsilon$ the energy of the incoming photon and E_u the Urbach energy. The E_U is estimated from the exponential part of the curve gave by equation 3. In the particular case of hydrogenated amorphous silicon the Urbach energy is interpreted as an indication of the film quality. Low values (around 50 meV) indicate a film with very low density of defects [13, 15].

In this work, the dependence of the optical coefficient on exposure to light and air have been studied by two experiments, exposure the sample on illumination conditions, and exposure to light and air conditions. The reason why to expose samples to light is to be searched in our necessity to force degradation in our films in order to be able to recognize its effects on optical spectra and distinguish them in other cases. In particular, changes in the low-absorption level and in the optical gap are two specific cases that will be checked. These experiments were carried on single and coevaporated organic semiconductor thin films.

EXPERIMENTAL DETAILS

Organic thin films have been deposited by thermal evaporation in a high vacuum chamber with a base pressure of 10^{-6} mbar with the substrate at room temperature. The powder sources were used without extra purification process. The substrates were Corning glass 7059 for optical and structural studies. The deposition rate was 3 Å/s and the thickness of the samples was 1 μm as measured by using a profilometer. In order to study the degradation process on samples exposed to light irradiation, a series of samples deposited in the same run were illuminated whereas others were left in normal enviromental conditions (labelled as "*n.c.*"). The illuminated samples were irradiated constantly for several periods of time. In this work we compared the as deposited optical parameters (Optical gap and Urbach energy) with the degrade samples. The structural properties were study by XRD technique. The light source used to degrade the samples was a halogen bulb, regulated at 1 sun power approximately, with an extra IR filter added in order to avoid heating in the films.

The optical transmission has been measured in the visible and near infrared range by means of an Ocean Optics spectrophotometer. For PDS measurements, the inert liquid (Fluorinert™ FC-40), in which the sample is submerged, was used. Measurements by PDS and transmittance have been repeated after the degradation processes and compared with the ones performed on as-deposited samples. In the cases in which no change has been found, the degradation process was prolonged and the measurements repeated, for this reason the days of degradation experiments are different. The XRD spectrum was taken with a diffractometer (Siemens D-500) in θ-2θ Bragg-Brentano geometry.

DISCUSSION

In Figure 1a, the PDS measurements show the effects of degradation on absorption coefficient of CuPc, where the typical Q and Soret energy bands can be seen [9]. The XRD spectra of CuPc thin film show that as deposited layer has polycrystalline structure (not show here). The value obtained for the optical gap is 1.64 eV, while for the Urbach energy a value of 53meV has been calculated .The Q energy band is also visible in the transmittance spectrum for values of wavelength in between 550 and 750 nm, (see Figure 1b). The irradiation with light causes an increase in Urbach energy; on the contrary, when the sample is exposed to environmental conditions, the variations of E_U are less systematic. The results are shown in Table 1. If we consider that the error affecting the results here presented is equal to ± 1 meV for E_U (and 0.01 eV for E_{og}) in all of the cases, such variations are not indicating a clear trend.

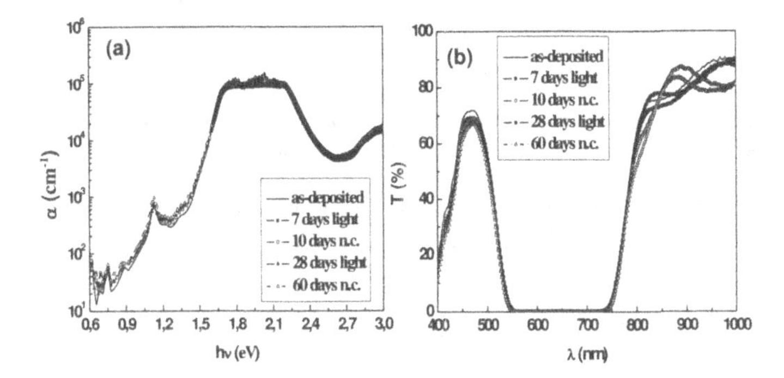

Figure 1: (a) PDS and (b) Transmittance measurements of CuPc layer exposure at light and air conditions.

The Figure 2a shows the PDS measurements of C_{60} single layer treated with the same processes of the CuPc. The as deposited layer shows no clear exponential region. The XRD spectrum shows that C_{60} has a big amorphous component, so the E_{og}, calculated by Tauc law, was 1.66 eV, in agreement with one reported by Gotoh *et al* [15] and the E_U was 63 meV. The absorption of the material in sub-gap region suffers from important changes either if exposed to direct illumination or if simply exposed to environmental air and temperature conditions. The treatment with light irradiation has stronger effects than on CuPc. As it can be seen in inset of Figure 2a, the Urbach tail is not completely clear, but an estimation of 140 meV has been obtained for the sample when exposed to irradiation with light, that indicate that the treatment makes the density of the band tail states to increase. When the material is simply exposed to normal environmental conditions of temperature and light the Urbach energy does not change and maintain the original value of 63 meV.

The PDS measurement of PTCDA is shown in Figure 2 (b). The polycrystalline, measured by XRD, thin film presents an optical gap of 2.11 eV and E_U of 46 meV. Only slight changes can be observed in the Urbach energy with the degradation processes: a very small increase proportional to the light irradiation exposure time. The process of degradation by exposure to environmental air and temperature has no evident effect.

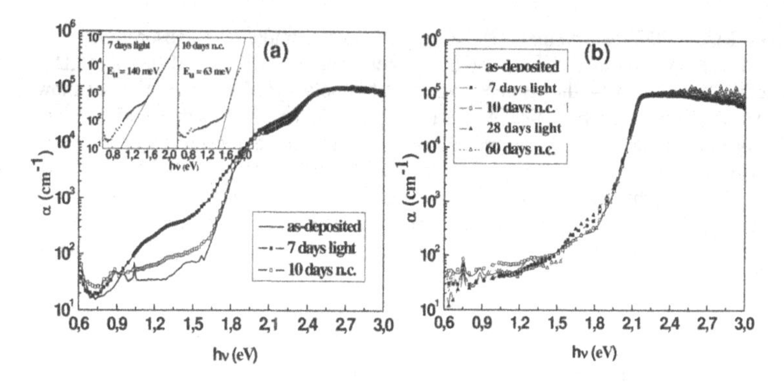

Figure 2: PDS measurements of (a) degraded C60 thin films and calculation of the Urbach energy (E_u), (on the vertical axis the scale is the same for both graphs); (b) degraded PTCDA thin-films.

Table I: The parameters obtained for single layers exposure to the degradation experiments.

		As dep.	Light irradiation ^		Air and Light ^	
			7	28	10	60
CuPc	E_{og}	1,64	1,64	1,64	1,64	1,64
	E_U	53	53	57	50	53
C60	E_{og}	1,66	1,65	-	1,66	-
	E_U	63	140	-	63	-
PTCDA	E_{og}	2,11	2,11	2,11	2,11	2,11
	E_U	46	48	50	47	47
*E_{og} in eV, E_U in meV. ^ Exposure days.						

Since organic solar cells are composed of acceptor and donor materials blended together to form a bulk heterojunction material, we have studied the optical absorption coefficients of combination of the previous p- and n-type materials: CuPc:C60 and CuPc:PTCDA. The film presents an amorphous structure so Tauc law corresponded to apply in order to calculated de optical gap. The PDS measurements of these co deposited layers are shown in Figure 3.

When Tauc law has been plotted in function of the energy two regions of the spectrum have been found in CuPc:C_{60} coevaporated sample: one between 1.65 and 1.78 eV and another one between 1.06 and 1.46 eV. The fit has been evaluated in both cases obtaining a result for the gap of 1.50 eV and 0.99 eV, respectively. The presence of two gaps resembles the case of indirect allowed transitions, where the creation or destruction of phonons is required during the process [16], but this is not expected in an amorphous layer. The energy of Urbach is equal to 77 meV, a value that is higher than both the ones of the pure materials. The obtained value is quite high for a good semiconductor but it could be expected for a co-evaporated thin-film, since the structural disorder should be higher. In Table II the changes in the optical gap and Urbach energy by degradation process are presented. It can be seen that both degradation processes yield an increase of absorption in sub-gap region, by an amount that is proportional to the time of exposure, but with not significant changes in E_{og}.

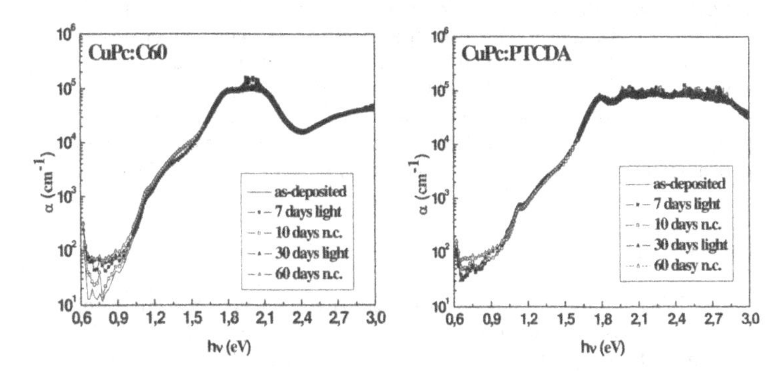

Figure 3: PDS measurements of degraded CuPc:C60 and CuPc:PTCDA coevaporated thin-films.

In the CuPc:PTCDA codeposited film, the degradation processes do not have any effect on the absorption coefficient. Again the optical gap seems to be low, if compared with the values of single layers. On the other side, the new absorptions found in the sub-gap region are responsible for the relevant changes of absorption spectrum and the result is also visible by naked eye, since the sample is characterized by a very obscure colour.

Table II. The parameters obtained for the bulk-heterojunctions exposure to the degradation experiments.

		As depos.	Light irradiation ^		Air and Light ^	
			7	30	10	60
CuPc: C60	$E_{og,1}$	0.99	0.99	0.98	0.99	1.03
	$E_{og,2}$	1.50	1.49	1.49	1.48	1.48
	E_U	77	83	90	78	88
CuPc: PTCDA	$E_{og,1}$	1.00	1.00	1.00	1.03	0.99
	$E_{og,2}$	1.52	1.51	1.50	1.52	1.51
	E_U	112	120	128	121	128
*E_{og} in eV, E_U in meV. ^ Exposure days.						

The effects of light degradation on the optical properties of acceptor, donor and acceptor/donor blends have been studied by Photothermal Deflection Spectroscopy. Moreover, the structure of the films has also studied by X-Ray diffraction spectroscopy.

CuPc and PTCDA evaporated thin-films present polycrystalline structure, whereas C_{60} thin-films show an amorphous phase. This amorphous phase is also present in the two coevaporated films (CuPc/C_{60} and PTCDA/C_{60}) studied in this work. The Urbach energy has been found higher for the amorphous materials than for polycrystalline ones.

C60 undergo much degradation when illuminated than under normal conditions, having their absorption coefficient increased a lot in the sub-gap region. Such result can be interpreted as caused by an increase in the density of states in the gap, probably associated with defects. The absorptions of CuPc and PTCDA suffer only from little modifications with both processes.

In the case of heterojunctions seem that CuPc reduce the degradation rate, especially in combination with high instable C60, because the changes observed in E_U is lower than the C_{60} single layer. Nevertheless the optical gap and the Urbach energy values for heterojunction samples must be taken just as a qualitative indication. It can be seen that the optical gaps of the heterojunctions are always lower than the ones calculated for the pure materials of which they are composed of.

ACKNOWLEDGEMENTS

We acknowledge financial support from the Consolider HOPE project CSD2007-07 of the Spanish government. Work has been partially supported from the European Community's 7th Framework Programme (FP7/2007-2013) under grant agreement n. 227127. We also thank the financial collaboration from XaRMAE of the Generalitat of Catalunya.

REFERENCES

1. F. Yang, M. Shtein, S.R. Forrest, *Nature Materials* **4**, p. 37 (2005).
2. A.A.M. Farag, *Optics & Laser Technology* **39**, p.728, (2007).
3.J.N.Haddock, X. Zhang, B. Domercq, B. Kippelen *Organic Electronics* **6**, 182-187 (2005) .
4. S. Uchida, J. Xue, B. P. Rand, S. R. Forrest, *Appl. Phys.Lett.***84**, 3013,(2004).
5. P. Peumans and S. R. Forrest, *App. Phys. Lett.* Vol. **79**, 126 (2001).
6. J.G. Xue, B.P. Rand, S. Uchida, S.R. Forrest, *Adv. Mater.* **17**, 66 (2005).
7. M.Stella, C. Voz, J. Puigdollers, F.Rojas, M. Fonrodona, J.Escarré, J.M. Asensi, J. Bertomeu, J. Andreu, *Journal of Non-Crystaline Solids* **352**, 1663, (2006).
8. C. Voz, J. Puigdollers, S. Cheylan, M. Fonrodona, M. Stella, J. Andreu, R. Alcubilla, *Thin Solids Films* **515**, 7675, (2007).
9. Y. Xahin, S. Alem, R. de Bettignies, J. Nunzi, *Thin Solid Films* **476**, 340 (2005) .
10. P.Peumans, S.R. Forrest, *Appl. Phys.Lett.***79**, 126,(2001).
11. K.V. Shalimova, *"Physics of Semiconductors"*, Ed. Mir (1975) p. 268.
12. M. M. El-Nahass, F.S. Bahabri and S.R. Al-Harbi, *Egypt. J. Sol.* **24**, 11-19 (2001) .
13. R.A. Street, *"Hydrogenated Amorphous Silicon"*, Cambridge Solid State Science Series (1991), p. 87.
14. R.A. Street, *"Hydrogenated Amorphous Silicon"*, Cambridge Solid State Science Series (1991), p. 88.
15. T. Gotoh, S. Nonomura, S. Hirata, S. Nitta, *Appl. Surf. Sci.* **113/114,** 278-281 (1997).
16. N. Karl, *Synthetic Metals* **133–134, p.**649–657 (2003).

Mater. Res. Soc. Symp. Proc. Vol. 1154 © 2009 Materials Research Society 1154-B10-55

Three-Dimensional Organic Field-Effect Transistors: Charge Accumulation in Their Vertical Semiconductor Channels

M. Uno[1, 2], I. Doi[3], K. Takimiya[3], and J. Takeya[1]
[1]Graduate School of Science, Osaka University, Toyonaka 560-0043, Japan
[2]Technology Research Institute of Osaka Pref., Izumi 594-1157, Japan
[3]Graduate School of Applied Chemistry, Hiroshima University 739-8527, Japan

ABSTRACT

Three-dimensional organic field-effect transistors are developed with multiple vertical channels of organic semiconductors to gain high output current and high on-off ratio. High-mobility and air-stable dinaphtho[2,3-b:2',3'-f]thieno[3,2-b]thiophene thin films deposited on horizontally elongated vertical sidewalls have realized unprecedented high output current per area of 2.6 A/cm^2 with the application of drain voltage -10 V and gate voltage -20 V. The on-off ratio is as high as 2.7x10^6. Carrier mobility of the organic semiconductor deposited on the vertical sidewalls is typically 0.30 cm^2/Vs. The structure is built also on plastic substrates, where still considerable current modulation is preserved with high output current per area of 70 mA/cm^2 and with high on-off ratio of 8.7x10^6. The performance exceeds practical requirements for applications in driving organic light-emitting diodes in active-matrix displays. The technique of gating with electric double layers of ionic liquid is also introduced to the three-dimensional transistor structure.

INTRODUCTION

Organic electronics have attracted much attention as "post-silicon electronics" due to their mechanical flexibility, low cost, energy-saving and environmental-friendly fabrication processes. Organic field-effect transistors (OFETs) are the key devices in which further development is necessary for such applications as full-flexible displays, incorporated in their matrix-controlling elements. So far, though much effort have been devoted to material development, even the best value of the carrier mobility μ of organic semiconductor films remain in the order of 1 cm^2/Vs [1-4], except for single-crystal devices [5-7]. Therefore, their performance in current amplification per pixel is not necessarily sufficient for practical use. Though decreasing the channel length L and increasing the width W can be another approach, this effect is limited if one cares the condition that the channels and the electrodes are in the same plane as the conventional OFETs. In this paper, we propose a three-dimensional organic field-effect transistor (3D-OFET) [8,9] to accumulate charge in its vertical semiconductor channels, so that space availability for the field-induced carriers is essentially enlarged.

EXPERIMENT

Figures 1(a) and 1(b) illustrates schematic cross-sectional views of conventional planar-type and 3D-type OFETs, respectively. In 3D-OFET, the same metal-insulator-semiconductor structure as the planar OFET is included, so that output current is modulated by the gate electric field with the identical mechanism but in the different direction. In Fig. 1(b), gate electric field is applied parallel to the substrate to accumulate carriers which flow from the source electrode to the drain electrode in the vertical channels.

Fabrication of 3D-OFETs on Si substrates

The structure of the 3D-OFET is fabricated with the following processes: a photoresist layer is spin-coated on a doped Si substrate and patterned to multi-block shape of 10 μm pitch in a 200 μm x 200 μm square area with the hight of approximately 3 μm. Then the Si substrate is dry-etched perpendicularly by SF_6 gas to the depth of 3 μm using the photoresist pattern for a mask. After removing the photoresist, 150-nm-thick SiO_2 is thermally grown to form a gate insulator. Organic semiconductor dinaphtho[2,3-b:2',3'-f] thieno[3,2-b]thiophene (DNTT) is vacuum deposited at room temperature on the vertical sidewalls of the multi blocks, after treatment of the SiO_2 surface with an octadecyltrichlorosilane self-assembled monolayer. Deposition of the DNTT films is performed from the diagonal direction of about 60 degree to the substrate to cover the whole sidewalls from the bottom to the top of the structure. The thickness of the films is around 50 nm. Finally, 10-nm-thick gold films are deposited from the strict upright direction to the structure to form both the source and the drain electrodes. Since the sidewalls of the multi-block structure are highly perpendicular to the substrate, gold films deposited on the land part and the groove part can be separated electrically to form both the drain and source electrode, respectively. In the resultant top-contact device, the channel width W corresponds to the total length of all the edges of the multi-column shape and channel length L equals to the height of the columns. Therefore, one can design the devices with huge ratio of W/L, which results in large enhancement of output current in response to the gate voltage, according to Eq.(1). In the present structure, the ratio of W/L equals to 910. Figure 1(c) shows overhead SEM images of the present 3D-OFET. The inset shows an expanded image of the multi blocks in the main panel.

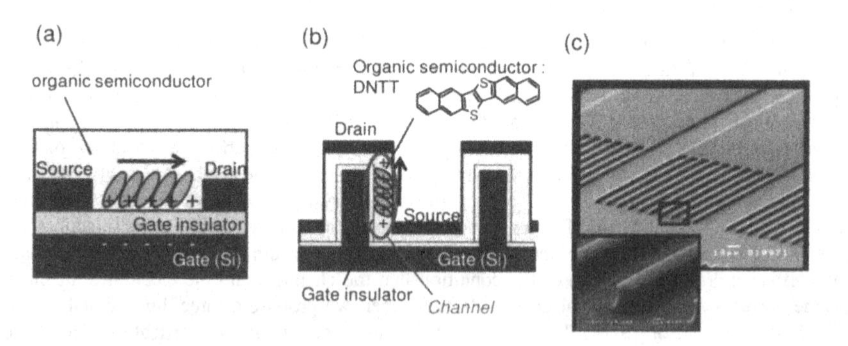

Figure 1. Illustrations of cross-sectional views of the typical structures of (a) conventional lateral OFET, and (b) 3D-OFET with molecular structure of the organic semiconductor DNTT. (c) SEM images of the 3D-OFET. The inset shows an expanded view of the micro columns.

3D-OFET on a plastic platform

On the way towards developing the 3D-OFETs for the organic flexible displays, the next challenge is to build them on plastic substrates. We fabricated the 3D structure in the following processes: the multi-columnar structure of the 3D-OFET is built by photolithography using SU8 epoxy photoresist (Kayaku Microchem Corp.) on a poly(ethylene naphthalate) (PEN) substrate. Then, Ti/Pt/Ti films are deposited as gate electrode by sputtering twice from the right- and left-top diagonal directions to cover the whole surface of the microstructure. A parylene film is deposited on the structure to the thickness of 500 nm to form a gate insulator, coating all the surfaces of the structure equally. Then, organic semiconductor DNTT is deposited from diagonal direction and gold films are deposited from upright direction in the same conditions as described previously. In the present experiment, the multi-block of SU8 is designed to 20 μm pitch and approximately 5 μm in the height in a 200 μm x 200 μm square-shaped area.

Gating with ionic liquids

With the aim to realize operation at much lower gate voltage, we also propose to use an electric double-layer (EDL) capacitance C_{EDL} using ionic liquid to accumulate carriers in the vertical channels [10], instead of using dielectric insulators. Gating with ionic liquids enables high-density charge accumulation up to 5×10^{13} cm^{-2} at very small gate voltages typically within 1.0 V. Moreover, the process of fabricating the liquid-to-semiconductor interface is relatively easy even for surfaces with such a complicated shape as of the 3D structure. For the present test devices, we simply filled the open surface of the organic semiconductor with the ionic liquid utilizing already fabricated for the 3D-OFETs shown in Fig. 1(a). Choosing the ionic liquid of *1*-ethyl-*3*-methyl-imidazorium bis(trifluoromethyl-sulfonyl)imid [emiTFSI] with relatively low viscosity, the liquid is introduced into the micro-columnar structure by itself. Figures 2(a) and 2(b) show a schematic cross-sectional view of the fabricated device and chemical formula of emiTFSI, respectively.

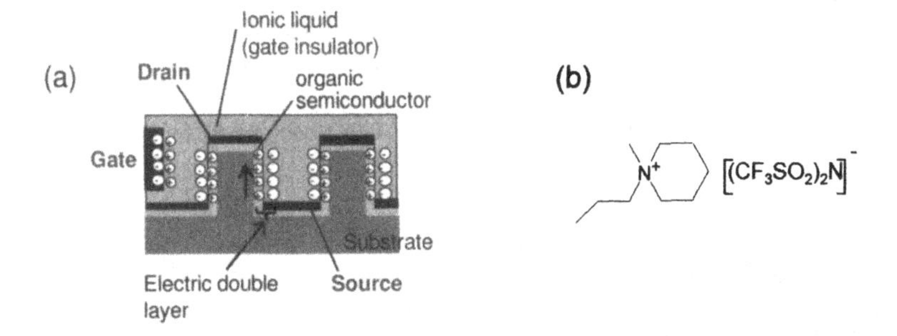

Figure 2. (a) Illustration of the cross-sectional structure when gating with ionic liquids.
(b) Chemical formula of the used ionic liquid, *1*-ethyl-*3*-methyl-imidazorium bis(trifluoromethyl-sulfonyl)imid [emiTFSI].

RESULTS AND DISCUSSION

Figures 3(a) and 3(b) show transfer and output characteristics of the 3D-OFET on the Si substrate, respectively. Pronounced field-effect modulation of the output current is observed in response to the applied gate voltage V_G, where I_D up to 2.6 A per cm^2 area (1.05 mA per 200 μm x 200 μm square area) is achieved with drain voltage V_D of -10 V and V_G of -20 V. It is to be emphasized that high on-off ratio of 2.7×10^6 is achieved, which is outstanding compared with other vertical OFETs fabricated [11-13]. Since typical requirement of output current for matrix-controlling elements in displays may be considered as about 1 μA per 100 μm x 100 μm square area [14], the present result exceeds the requirement by more than two orders of magnitude. Carrier mobility μ of the DNTT film in the vertical channel is estimated to be typically 0.3 cm^2/Vs, which is still less than the typical values of 1.2 cm^2/V s for lateral devices prepared on the same substrate as a reference.

We note that I_D does not exhibit saturating trend in Fig. 3(b) even with $|V_D|$ larger than $|V_G|$. Since the thickness (150 nm) of the gate insulator is not negligible in the present devices as compared to the channel length (3.0 μm), the short-channel effect may be involved in the characteristics [15], with the pinch-off region not necessarily confined in the vicinity of the contact to the drain electrode. Moreover, the gate electric field would not be ideally applied in the channel region at the distance less than the insulator thickness from the top drain electrode, as speculated on Fig. 1(b). This effect could elongate the pinch-off region so that the current saturation is further reduced in the output characteristics. It is therefore crucial to develop the devices with essentially thinner gate insulator, which is currently undergoing in our laboratory.

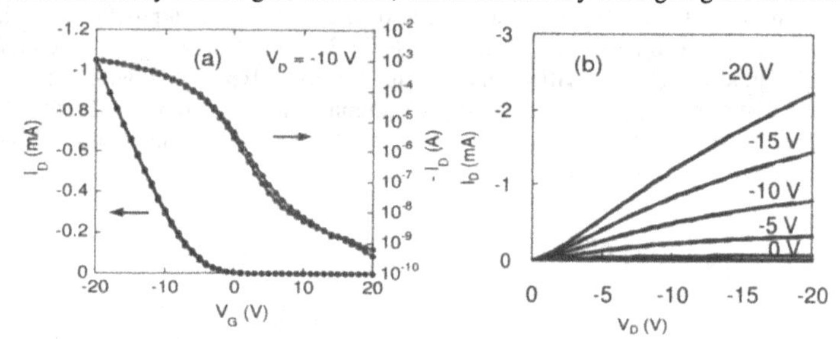

Figure 3. (a) Transfer and (b) output characteristics of 3D-OFET on Si substrate.

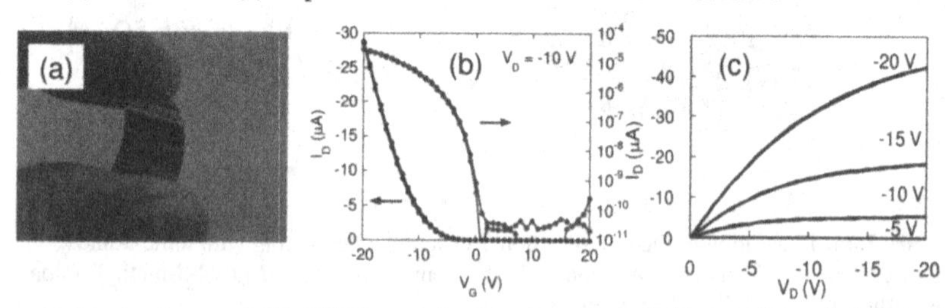

Figure 4. (a) A picture of 3D-OFETs on plastic substrate, and (b) transfer and (c) output characteristics of the 3D device on plastic substrate.

Figure 4(a) shows a picture of a 3D-OFET fabricated on a plastic substrate. Figures 4(b) and (c) show transfer and output characteristics of the 3D device, respectively. In Fig. 4(b), high output current up to 70 mA per cm^2 area (28 μA per 200 μm x 200 μm square area) is achieved with V_D of -10 V and V_G of -20 V. The on-off ratio of 8.7×10^6 is even higher than the value for the previous device on the Si/SiO_2 substrate. For thcase present PEN devices, carrier mobility μ of the DNTT film in the vertical channel is estimated to be typically 0.1 cm^2/Vs, which is less than that on SiO_2/Si substrate shown in Fig. 1(b), possibly because of poorer alignment of the DNTT molecules. In Fig. 4(b), subthreshold swing is improved compared with the 3D-OFET on the Si substrate, perhaps because the sidewalls of SU8 resist can be formed strictly vertical to the substrate and the uniformity of the applied gate electric field is improved. The results already demonstrate usefulness of the 3D structure in achieving sufficient current per pixel for matrix-controlling elements, which can drive industrial development of organic flexible displays. We note that the variation in the on-currents for the PEN/parylene device and the previous Si/SiO_2 device attributes mainly to the difference in gate dielectric thickness, meaning that further improvement in the performance is possible by merely reducing the thickness of the parylene.

For the prototype 3D-OFET with the ionic liquid, transfer and output characteristics are plotted in Figs. 5(a) and (b). Reasonable transistor characteristics are obtained for the ionic-liquid gated 3D-OFETs devices; 0.5 V is enough for both V_G and V_D to obtain I_D of 2 μA per 100 μm square area, resulted from much larger C_{EDL} but smaller μ, which is typically 1×10^{-3} cm^2/Vs. Although there are plenty of rooms to improve the mobility at the surface of the organic semiconductor film, the technique of gating by electric double layers, which is a powerful tool to accumulate high-density charge at surface of organic semiconductors, can also be easily applied to 3D-OFET structure using the capillary force.

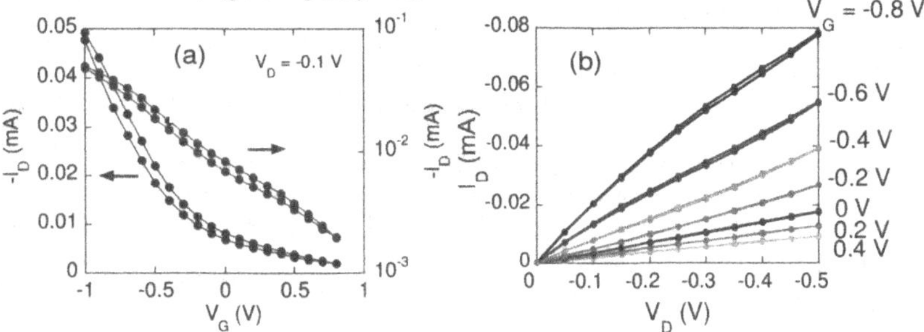

Figure 5. (a) Transfer and (b) output characteristics of 3D-OFET when gating with ionic liquid.

CONCLUSIONS

We have developed a 3D-OFET with a series of multiplied vertical semiconductor channels, enabling accumulation of unprecedentedly large amount of carriers per unit device area. Establishing a technique of forming high mobility organic semiconductors to the multiple vertical channels, sufficiently high output current and high on/off ratio are realized to enable practical applications such as matrix-controlling elements for OLED displays. For the devices fabricated on Si substrate, the output current of 0.26 mA per 100 μm x 100 μm square area is obtained for applied voltages of V_D of -10 V and V_G of -20 V, which is practical for circuitry power source. The on-off switching ratio is achieved more than 10^6 to guarantee high contrast for pixels. Placing 3D-OFETs on plastic platforms, similar enhancement of output current is achieved and high output current of 7 μA per 100 μm x 100 μm square area and high on-off ratio of 8.7×10^6 are obtained. Novel technique of gating with ionic liquid is also applied to 3D-OFET structure and is proven to operate to modulate the amount of accumulated charges with gate voltage. These results will make the real profit of the technology with the revolutionary high-performance flexible OFETs.

ACKNOWLEDGMENTS

This work was financially supported by Industrial Technology Research Grant Program in 2006 from NEDO, Japan, and by a Grant-in-Aid for Scientific Research (Nos. 17069003, 18028029, and 19360009) from MEXT, Japan.

REFERENCES

1. Y. Y. Lin, D. J. Gundlach, S. F. Nelson, and T. N. Jackson, IEEE Electron Device Lett. **18**, 606 (1997).
2. H. Klauk, M. Halik, U. Zschieschang, G. Schmid, W. Radlik, and W. Weber, J. Appl. Phys. **92**, 5259 (2002).
3. J. E. Anthony, J. S. Brooks, D. L. Eaton, and S. R. Parkin, J. Am. Chem. Soc. **123**, 9482 (2001).
4. T. Yamamoto and K. Takimiya, J. Am. Chem. Soc. **129**, 2224 (2007).
5. V. Podzorov, E. Menard, A. Borissov, V. Kiryukhin, J. A. Rogers, and M. E. Gershenson, Phys. Rev. Lett. **93**, 086602 (2004).
6. J. Takeya, M. Yamagishi, Y. Tominari, R. Hirahara, Y. Nakazawa, T. Nishikawa, T. Kawase, and T. Shimoda, Appl. Phys. Lett. **90**, 102120 (2007).
7. O. D. Jurchescu, M. Popinciuc, B. J. van Wees, and T. T. M. Palstra, Adv. Mater. (Weinheim, Ger.) **19**, 688 (2007).
8. M. Uno, Y. Tominari, and J. Takeya, Appl. Phys. Lett. **93**, 173301 (2008).
9. M. Uno, I. Doi, K. Takimiya, and J. Takeya, Appl. Phys. Lett. **94**, 103307 (2009).
10. S. Ono, S. Seki, R. Hirahara, Y. Tominari, and J. Takeya, Appl. Phys. Lett. **92**, 103313 (2008).
11. K. Kudo, D. X. Wang, M. Iizuka, S. Kuniyoshi, and K. Tanaka, Thin Solid Films **331**, 51 (1998).
12. R. Parashkov, E. Becker, S. Hartmann, G. Ginev, D. Schneider, H. Krautwald, T. Dobbertin, D. Metzdorf, F. Brunetti, C. Schildknecht, A. Kammoun, M. Brandes, T. Riedl, H.-H. Jo-hannes, and W. Kowalsky, Appl. Phys. Lett. **82**, 4579 (2003).
13. H. Naruse, S. Naka, and H. Okada, Appl. Phys. Express **1**, 011801 (2008).
14. Here, the requirement for the on current is given by the assumption that each organic light-emitting diode (OLED) with a certain bright area should be driven by the OFET that occupy *at most the same area*. Since OLEDs are to be driven in the condition of 3-6 cd/A for sufficient lifetime, the current for achieving usual brightness of 200-1000 cd/m^2 is calculated to be typically ~ 100 A/m^2, which equals to 10 mA/cm^2 and 1 μA per 100 μm x 100 μm area. See for example, K. M. Vaeth, Information Display **19**, 12 (2003).
15. J. N. Haddock, X. Zhang, S. Zheng, Q. Zhang, S. R. Marder, and B. Kippelen, Org. Electron. **7**, 45 (2006).

Mater. Res. Soc. Symp. Proc. Vol. 1154 © 2009 Materials Research Society 1154-B10-56

Reduced Contact Resistances in Organic Transistors With Secondary Gates on Source and Drain Electrodes

K. Nakayama[1], T. Uemura[1], M. Uno[1,2], and J. Takeya[1]
[1]Graduate School of Science, Osaka University,
1-1 Machikaneyama, Toyonaka 560-0043, Japan
[2]TRI-Osaka, Ayumino, Izumi 594-1157, Japan

ABSTRACT

Secondary-gate electrodes are introduced in organic thin-film transistors to reduce carrier-injection barriers into air-stable organic semiconductors. The additional gate electrodes buried in the gate insulators under the source and drain electrodes form "carrier-rich regions" in the vicinity of source and drain electrodes with the application of sufficiently high local electric fields. Fabricating the structure with dinaphtho[2,3-b:2',3'-f]thieno[3,2-b]thiophene, known for its excellent air-stability, it turned out that the contact resistance is drastically reduced especially when operated at low gate voltage in the main channel. The result demonstrates carrier injection with a minimized potential barrier realizing that from the *same* semiconductor material in the absence of peculiar interfacial trap levels at metal-to-semiconductor junctions.

INTRODUCTION

To realize the maximum device performance of organic field-effect transistors (OFETs), a significant challenge is to achieve efficient carrier injection from the electrodes. Such contact performance is the most seriously concerned for short-channel devices with the channel length typically less than sub-micrometers, though they are highly attractive because of their high-frequency response and capability of high-density integration. The problem of the contact barriers appear to be serious also for *air-stable p*-type organic semiconductors because stability to the oxidization is naively linked to difficulty to extract electrons causing slightly larger energy-level mismatch between the potentials of their highest occupied molecular orbitals (HOMOs) and the Fermi level of gold in most materials. Indeed, recently synthesized high-performance materials more air-stable than pentacene are reported to have HOMOs lower than that of pentacene by a few hundred mV [1,2]. Though contact resistance is reduced for such materials when relatively high gate voltage is applied and high-density carriers are accumulated in the channel, it is desired to develop methods of actively improving the contact problem in order to maximize the attractiveness of the high-mobility air-stable organic transistors with the capability of their low-power operation.

As compared to common silicon metal-oxide-semiconductor field-effect transistors (Si-MOSFET) where heavily doped carrier-rich region is incorporated next to the channel, all the reported organic field-effect transistors suffer from the carrier injection from different materials such as metals. Previously, we reported a device structure with "split gates" on the source and drain electrodes buried in the gate-insulating layers, so that the carrier density in the organic semiconductors in the vicinity of the source and drain electrodes can be varied independently of the primary gate electric fields applied to the central channel in the semiconductors [3]. Thereby, the carrier reservoirs are formed in the semiconductor locally near the electrodes by the secondary gate voltage. In this work, we have adopted the "split-gate" construction for an air-stable thin-film transistor of dinaphtho[2,3-b:2',3'-f]thieno[3,2-b]thiophene (DNTT) to study the carrier injection from the same semiconductor material, mimicking the case for the Si-MOSFETs. As the result of systematic measurements using the devices with different channel length, it is demonstrated that the carrier-rich regions under the split-gates give very small resistance for the hole injection that is negligible within the measurement accuracy.

EXPERIMENTAL

Device fabrication

To fabricate the DNTT transistors with the secondary gates for the carrier injection from the same material, we employ the structure that we described in our previous report [3]. Figure 1(a) shows the device structure with the split electrodes for the secondary gating buried under the source and drain electrodes, so that the carrier-rich regions are locally formed near the source and drain electrodes by applying sufficiently high voltage to the split-gate electrodes.

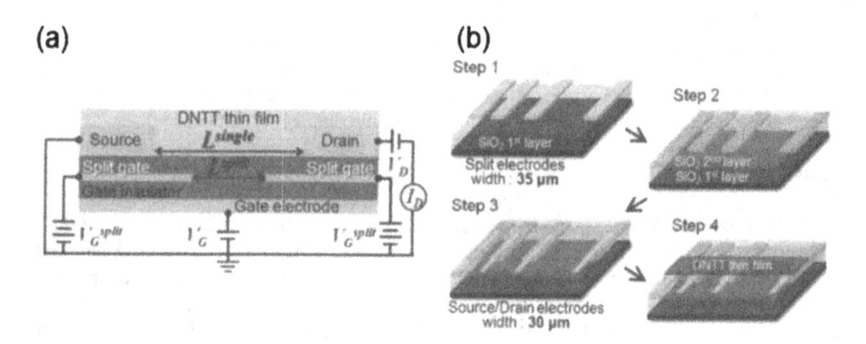

Figure 1. (a) Schematic illustration of the split-gate organic transistor. (b) Fabrication process of the split-gate DNTT transistor for the gated-transmission line measurement.

Shown in Fig. 1(b) is the fabrication process of the split-gate devices. We prepared the electrode patterns so that a series of devices with different channel lengths are fabricated at once in the same condition to evaluate contact resistance R_c using the gated-transmission line method (g-TLM) [4]. First, patterns for the split gate electrodes are drawn on 500-nm-thick SiO_2/doped silicon substrates to the width of 35 µm by photolithography. Then, 3-nm-thick titanium, 20-nm-thick platinum and 3-nm-thick titanium films are vacuum deposited consecutively. Thus obtained substrates are coated with 350-nm-thick SiO_2 films for the secondary gate dielectric layers by sputtering. The source and drain electrodes are formed with vacuum deposited Cr/Au films by photolithography to size of 30 µm slightly narrower than the above prepared split-gate electrodes as shown in Figs. 1(a) and (b). Thicknesses of Cr and Au are again 3 nm and 20 nm, respectively. Finally DNTT thin films are deposited to the thickness of 50 nm through a shadow mask at the rate of 0.5 Å/s. The channel length ranges from 25 μm to 400 μm and the channel width W is fixed to 300 μm.

Measurement method

The transistor performance is measured with three source measure units equipped in a Keithley 4200 semiconductor parameter analyzer; one applies the source-drain voltage V_D and measures the drain current I_D, another applies the primary gate voltage V_G to the doped Si substrate, and the other applies the split gate voltage V_G^{split} to the split-gate electrodes. To detect the influence of the split gates, we compare results of the following two operation modes.

(1) *Operation* of the *single-gate FET mode*

In the single-gate FET mode, both V_G and V_G^{split} are swept simultaneously so that

$$V_G^{split} = \frac{C_{total}}{C_{2nd}} V_G^{single} \tag{1}$$

is satisfied, where C_{total} and C_{2nd} are capacitances of the SiO_2 per area for the total thickness and for only the second layer, respectively. In this mode, the gate electric field is identical in the whole region of length L^{single} between the gate electric field. Therefore, the device operates in the same way as an ordinary bottom contact OFETs, in which carriers are introduced from the metallic electrode into the semiconductor channel with the length of L^{single}.

(2) *Operation* of the *split-gate FET mode*

In the split-gate FET mode, sufficiently high voltage is kept applied to the split gate electrodes to form the carrier-rich regions in the vicinity of the source and drain electrodes so that $|V_G^{split}| >> |V_G|$. Even with modifying V_G, the carrier density in the regions near the source and drain electrodes are constant and the hole-rich regions can be regarded as the electrodes of the *same* organic semiconductor material. As the result, the channel length is modified to L^{split}.

RESULTS AND DISCUSSION

Plotted in Fig. 2(a) are the output characteristics of one of the DNTT thin film transistors both in the single-gate mode and in the split-gate mode. L^{single} and L^{split} are 300 μm and 295 μm, respectively. In the split gate mode, V_G^{split} is fixed to -100 V. Since the main gate voltage V_G is only -10 V, the condition that $|V_G^{split}| >> |V_G|$ is fulfilled.

Drain current I_D is larger in the split-gate mode than that in the split-gate mode in the whole range of drain voltage V_D. Since the effective channel lengths of L^{single} and L^{split} are almost identical the difference in I_D is mostly attributed to the difference in the contact resistance R_c in the two operation modes, demonstrating the improvement in the carrier injection from the carrier-rich region induced by finite V_G^{split}. We note that slightly nonlinear concave curvature is observed for the single-gate measurement with $|V_D| < 5$ V, while the feature is less pronounced for the split-gate measurement. The result also indicates that the carrier injection rate is improved due to the presence of the hole-rich region around the metal electrodes.

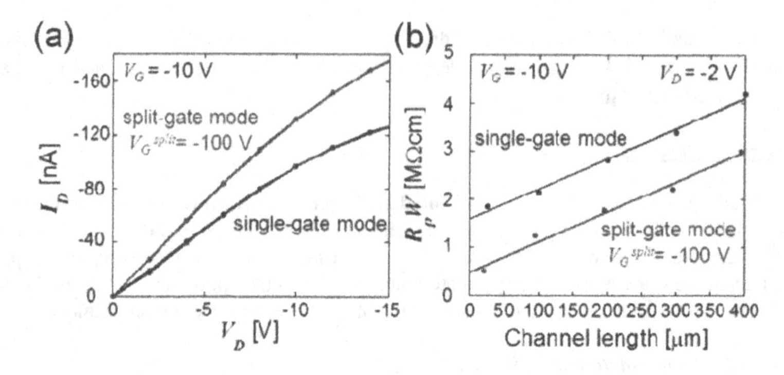

Figure 2. (a) Output characteristics of a DNTT organic transistor with the secondary gates. (b) Normalized resistance as function of channel length for a series of DNTT organic transistors with the secondary gates. The curves and lines in red and in blue are those for the split-gate mode and for the single-gate mode, respectively.

Following a standard technique of the g-TLM, we plot normalized total resistance R_pW of the devices as a function of the channel length L in both modes for a series of the DNTT thin-film transistors with the secondary gates. The total resistance of each device is calculated from nearly linear slope in the low V_D region of the I_D vs. V_D characteristics [see Fig. 1(a), for example]. Assuming that the total resistance is given as a series connection of the contact resistance R_c and the channel resistance, the L dependence of R_pW is to be represented by the equation

$$R_pW = R_cW + \frac{L}{\mu C_{total}(V_G - V_{TH})} \quad (2),$$

where μ, C_{total} and V_{TH} represents channel mobility (free from the contact resistances), the total capacitance and threshold voltage. Therefore, R_cW is evaluated as the intercept of the line fitted to the data of R_pW vs. L, as shown in Fig. 2(b). L can be L^{single} and L^{split} for the single-gate and split-gate modes, respectively.

One can first notice that the results in Fig. 2(b) are well described by the g-TLM from the fact that the R_cW - L data are approximately aligned in lines for the both modes. Moreover, since the contact-free mobility is extracted from the slope of the line of the R_pW vs. L plot, the result of nearly identical slopes of the two lines for the single- and split-gate modes also guarantees the validity of the model. Therefore, the quality of the DNTT thin films is homogeneous over the whole regions and all the contacts can be regarded as identical. The channel mobility μ of the film is estimated to be ~ 0.13 cm^2/Vs.

Following the above argument, the values of R_cW in the single- and split-gate modes are estimated to be 0.5 MΩcm and 1.6 MΩcm from the intercept, respectively. The result of the reduced value of R_cW to one third for the split-gate mode is attributed to the effect of the improvement in the carrier injection rate from the hole-rich region of the DNTT film.

Dependence of the contact resistance on the gate voltage is derived from transfer characteristics and is shown in Fig. 3. In the single gate mode, R_cW strongly depends upon the gate voltage with rapid enhancement in the low-voltage region. In contrast, R_cW in the split gate mode are not much dependent on V_G. The results demonstrate that the effect of the reduced injection rate from the hole-rich region DNTT is more pronounced in the region of lower gate voltages.

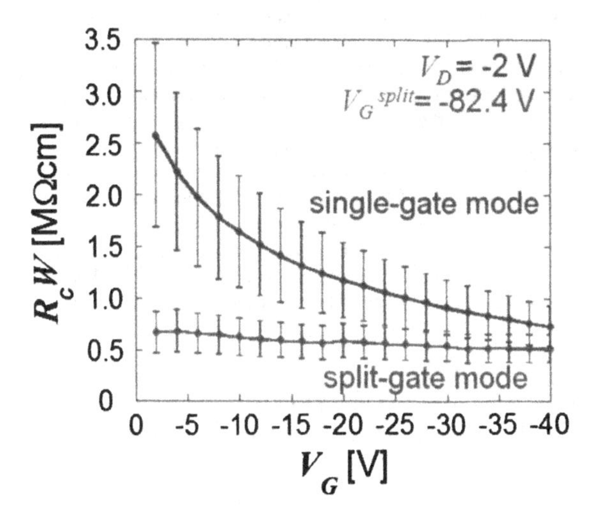

Figure 3. Normalized contact resistance as a function of gate voltage for a series of DNTT organic transistors with the secondary gates. The curves in red and in blue are for the split-gate mode and for the single-gate mode, respectively.

For the structure shown in Fig. 1(a), it can be argued that the normalized resistance in the split-gate mode is composed of the contact resistance $R_{Au\text{-}DNTT}W$ at the Au-to-DNTT interface, resistance of the "hole-rich" regions with the total length of 5 μm that are located on the split gates, and resistance for the injection from the "hole-rich" region. In the measurement condition of sufficiently large V_G^{split}, the resistance of the "hole-rich" regions is negligible. Noting that further application V_G of *in the measurement of the single-gate mode* should give an estimation of the contact resistance at the Au-to-DNTT interface *in the measurement of the split-gate mode*, one can have an idea that $R_{Au\text{-}DNTT}W$ forms realtively large part in the total R_cW in the split-gate mode. As the result, the portion of the net resistance for the injection from the "hole-rich" region is likely to be small in the total R_cW. Since the injection rate differs with charge accumulation in the semiconductor channel, the observation of almost V_G independent is also consistent with the picture. Though realistic data at higher gate voltage than shown in Fig. 3 are required for more quantitative separation, our measurement is not successful in such a region because the present devices suffer from deviation from the behavior of Eq. (2) due to bias-stress effect which is small but different among the devices on the same substrates. Improvement in the fabrication processes and/or gate-insulating material is necessary.

CONCLUSIONS

The effect of the secondary gate electrodes for inducing "hole-rich" region in the vicinity of source and drain electrodes is demonstrated in air-stable organic thin-film transistors of DNTT, somewhat mimicking the ion-implanted carrier-injection domains in Si-MOSFETs. The contact resistances are estimated performing the gated-transmission line methods as a function of applied channel gate voltages for both the single-gate and split-gate measurements, so that the presence of "hole-rich" is highlighted by the direct comparison between the two for the identical semiconductor channels. The influence is most drastic in the region of low-gate voltages, which helps improving the performance of the air-stable OFETs that may suffer from relatively large injection barriers due to the tolerance to oxidization. To replace the application of high gate voltages to the split-gate electrodes, possible techniques such as depositing acceptor species or using ferroelectric gate insulators are promising to apply the method of inducing "hole-rich" regions to practical low-power devices to reduce contact resistances.

ACKNOWLEDGMENTS

This work was financially supported by Industrial Technology Research Grant Program in 2006 from NEDO, Japan, and by a Grant-in-Aid for Scientific Research (Nos. 17069003 and 19360009) from MEXT, Japan.

REFERENCES

1. H. Ebata, T. Izawa, E. Miyazaki, K. Takimiya, M. Ikeda, H.Kuwabara, and T. Yui, J. Am. Chem. Soc. **129**, 15732 (2007).
2. N. Kawasaki, Y. Kubozono, H. Okamoto, A. Fujiwara, and M. Yamaji, Appl. Phys. Lett. **94**, 043310 (2009).
3. K. Nakayama, K. Hara, Y. Tominari, M. Yamagishi, and J. Takeya Appl. Phys. Lett. **93**, 153302 (2008).
4. J. Zaumseil, K. W. Baldwin, and J. A. Rogers, J. Appl. Phys. **93**, 6117 (2003).

Mater. Res. Soc. Symp. Proc. Vol. 1154 © 2009 Materials Research Society 1154-B10-59

Low-Temperature Thermal Conductivity of Rubrene Single Crystals: Quantitative Estimation of Defect Density in Bulk and Film Crystals

Y. Okada[1], M. Uno[1,2], and J. Takeya[1,3]
[1]Graduate School of Science, Osaka University,
1-1 Machikaneyama, Toyonaka 560-0043, Japan
[2]TRI-Osaka, Ayumino, Izumi 594-1157, Japan
[3]PRESTO-JST, Kawaguchi 333-0012, Japan

ABSTRACT

Thermal conductivity of rubrene single crystals is measured for both bulk and film-like crystals down to 0.5 K in order to estimate quantitatively density of crystalline defects through their phonon mean free paths. The temperature profile of the bulk rubrene crystals exhibit pronounced peak at ~ 10 K in the thermal conductivity as the result of very long mean-free paths of their phonons which indicates extremely low-level defect density in the region of 10^{15}-10^{16} cm^{-3} depending on different growth methods. The crystals grown from gas phase tend to have less defects than those grown from solution. It turned out that the film-like crystals have a few times more defect density as the result of the measurement by using newly developed devices for minute crystals.

INTRODUCTION

Recently, organic crystals are gaining considerable interest linked to practical applications as electronic semiconductor devices and nonlinear optical components; rubrene single crystal field-effect transistors exhibit one-order higher performance as compared to popular polycrystalline thin-film organic transistors and such materials as DAST shows excellent nonlinearity in response to light irradiation because of vibronic coupling to molecular polarization. However, it is concerned that density of dilutely distributed defects in these crystals is generally difficult to define, though it is recognized as crucial factor for transport or optical performances of the above devices. In this work, we measured low-temperature thermal conductivity of phonons to estimate their mean-free paths and density of crystalline defects responsible for scattering the phonons. It is known that temperature dependence of thermal conductivity in high-quality single crystals shows a pronounced peak in the temperature range below typically 10-30 K, below which phonons are predominantly scattered by defects. At higher temperatures, phonon-phonon Umklapp process dominates their scattering events.

EXPERIMENTAL

Crystal grwoth

Bulk rubrene single crystals are grown both from gas phase by physical vapor transport [1-3] and from solution [4]. Commercial sublimed grade product of Aldrich 551112 was used for the starting material. For the gas-phase growth, the raw materials are vaporized at 295 °C and crystallized at 240 °C in a two-zone tube furnace in a stream of Ar gas using the technique of Physical Vaport Transport (PVT) [1]. The resultant crystals were brought to the sublimation boat and were revaporized so that purer rubrene crystals can be grown. We note that a small amount of differently colored powders appeared as the result of the first growth, whereas these apparent impurity compounds did not show up after the second crystal growth. Bulk crystals are obtained in the condition of slow growth with weak gas flow, whereas thin-film-like crystals with typical thickness of 1 μm are obtained in rapidly in a day with faster gas flow. We reported that very high-mobility organic single-crystal transistors were realized with the latter rubrene crystals [5].

For the growth from the solution, rubrene solutions with aniline and p-xylene were prepared so that it had a saturation point of 45 °C. The solutions were set in a temperature-controlled water bath and heated up to 60 °C in dark. Using a program of temperature control, the solutions were first cooled to 45 °C and further cooled in an extremely slow rate. It usually takes a few weeks before growing mm-size crystals.

Methods of thermal conductivity measurement

Thermal conductivity κ is measured from 0.5 K to room temperature using a steady-state one heater two thermometer technique. As shown in Fig. 1(a), a set of mm-size chip resistors are mechanically attached for stand-alone bulk crystals in case they have the length of subcentimeter. κ is estimated as the coefficient of the linear response of temperature gradient to the continuous heat flow Q, which is normalized to heat flow density Q/S with the cross-section S. Since the temperature gradient is measured by temperature difference ΔT at the two positions on the samples and the distance L, κ is given as

$$\kappa = \frac{Q}{\Delta T}\frac{L}{S} \qquad (1).$$

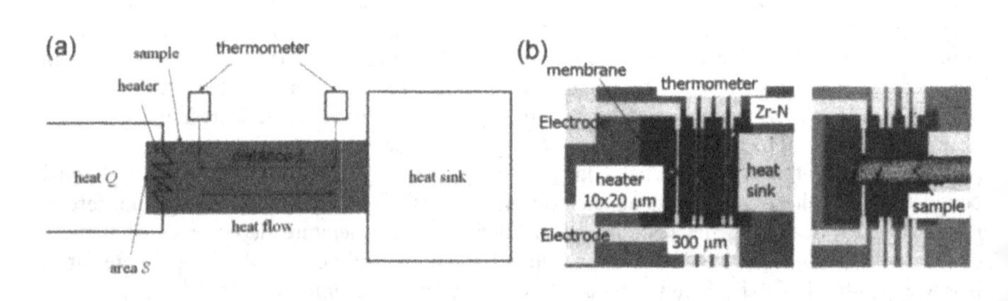

Figure 1. (a) A schematic diagram of the thermal conductivity measurement for bulk organic crystals. (b) A newly developed MEMS device for thermal conductivity measurement of submillimeter organic crystals.

In addition to the stand-alone measurement, we have developed a chip-device incorporating resistive thin films of ZrN with the dimension of 100 μm, which work both for the heater and the thermometers. Employing MEMS (Micro Electro-Mechanical Systems) technique, these resistive thin films are sustained only by 1-μm thick membrane, so that the heat current is essentially restricted to the crystal. Using the latter technique, we can measure crystals of submillimeter sizes.

RESULTS AND DISCUSSION

Thermal conductivity of bulk rubrene crystals

Figure 2 shows temperature profiles of three bulk rubrene crystals grown by different techniques; Sample A is grown by PVT and the others are from solutions. Sample B and C are crystallized from aniline and p-xylene solutions, respectively. Typical dimensions of all the crystals are 1 mm in the direction of heat flow and 0.5 x 0.5 mm^2 for the cross-sections, respectively. The sample dependence is the most pronounced in the temperature range below ~ 30 K because of different phonon mean-free paths in the crystals. On the other hand, phonon-to-phonon scattering is dominant at higher temperatures, so that sample dependence is diminished in κ.

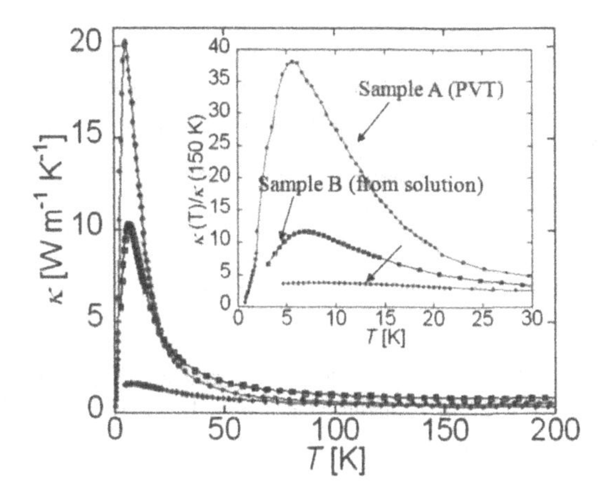

Figure 2. Thermal conductivity of three bulk rubrene crystals grown by different techniques. Sample A is by PVT, while B and C are from aniline and p-xylene solutions, respectively.

In order to have more quantitative estimate of the phonon mean-free paths l of the crystals, low-temperature plots are shown in Fig. 3(a) for Samples A and B because l is directly related to the defect densities. Since specific heat C is needed to estimate l from κ, the quantity is directly measured down to 0.4 K by Physical Properties Measurement System (Quantum Design

Co.). As shown in Fig. 3(b), the result gives typical temperature dependence proportional to T^3, given by the standard Debye model at low temperatures, i.e.

$$C = \beta T^3 \quad , \quad \beta = \frac{12\pi^4}{5} R \frac{1}{\theta^3} \qquad (2),$$

where θ = 254.5K is the Debye temperature and R is a coefficient given by the phonon frequency of the crystal. Therefore, l is determined with κ and C because

$$\kappa = \frac{1}{3} Cvl \qquad (3),$$

where v is the phonon velocity. Note that the specific heat is not dependent on samples because it is determined only by static material parameters.

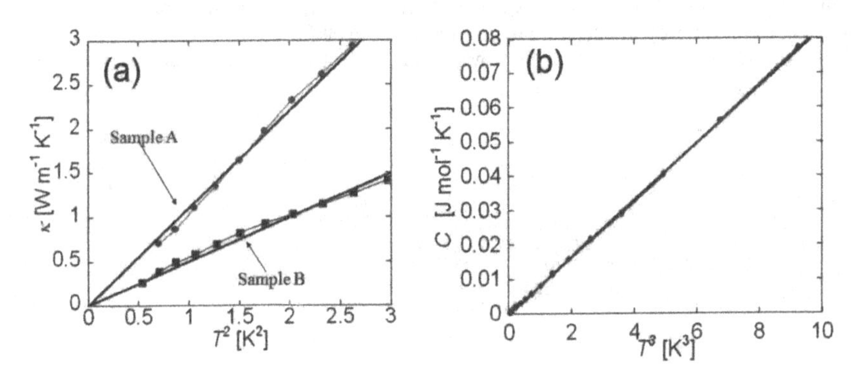

Figure 3. (a) Thermal conductivity of bulk rubrene crystals Sample A and B at low temperatures. (b) Specific heat of Sample A.

The temperature dependence in Fig. 3(a) clearly exhibits κ in proportion to T^2 indicating that the scattering source is mainly crystalline dislocations. Therefore, l is described as

$$l = \frac{1}{N_D} \frac{2\pi v}{\gamma^2 B^2 \omega} \qquad (4)$$

with the quantities of density N_D of the dislocations defect, Grueneisen constant g, the Berger's vector B in the order of the inter-lattice distance, and the dominant phonon frequency ω, which is typically 1/3 of the temperature [6]. After all, N_D is estimated as ~ 10^{15} cm^{-3} for Sample A and ~ 2 x 10^{15} cm^{-3} for Sample B, respectively. It is because of such small defect density that the rubrene crystals show the pronounced phonon peaks in the temperature profiles of thermal conductivity, which is rarely reported for organic crystals. Since Sample C does not show thermal conductivity peak, the defect density is much higher. Since Sample C does not show thermal conductivity peak, the defect density is much higher, indicating that the density of the strain dislocations is significantly dependent on the solvents for crystals grown from solution.

Thermal conductivity of film-like rubrene crystals

Figure 4 shows thermal conductivity of a submillimeter film-like crystal which is used for high-performance organic crystal transistors measured by using newly developed MEMS devices. A convenient method to examine the validity of the measurement is to compare the high-temperature side of the temperature profile because the value is governed by the phonon-phonon Umklap scattering independent of defect densities. As compared with Fig. 2, the high-temperature tail shows almost identical vales to the data for the bulk crystals, guaranteeing that the thermal conductivity is properly measured also for the thin-film crystal with the MEMS devices.

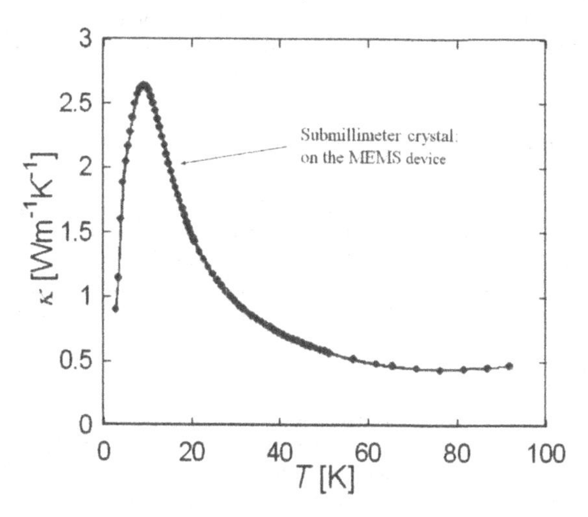

Figure 4. Temperature dependence of thermal conductivity in a film-like rubrene crystal measure using the developed MEMS device.

In the low-temperature region, pronounced phonon peak appears also in the film crystal, indicating relatively low defect density there. The density N_D can be roughly estimated to be ~ 5 x 10^{15} cm^{-3}. The cleanliness of the rubrene crystals may be linked to the high mobility achieved when laminated on gate-insulating substrates [5]. It is intriguing to address that the mobility values are somewhat related to the quality of the crystals evaluated in the thermal conductivity experiment. While the crystals grown from aniline show 1-3 cm^2/Vs [4] and gas-phase grown crystals usually provides even higher mobility [2,5] both for bulk and thin-film, the crystals grown from p-xylene solution did not result in clear transistor performance. The result is consistent with that only Sample C does not show thermal conductivity peak because of apparently higher density of dislocations. Further collections of comprehensive measurements of thermal conductivity and transistor characteristics are necessary to give more detailed connection from the crystalline quality to the mobility of the field-induced carriers and to separate effects of scattering or trapping potentials at the surface and relevance of the scattering in the bulk defects in the high-performance crystal devices.

CONCLUSIONS

The measurement of thermal conductivity in rubrene single crystals indeed has given qualitative and quantitative understanding about their crystalline defects. Qualitatively, it turned out that the dominant defects in the crystals are dislocations probably induced during the process of the slow crystal growth. Therefore, further slowing-down in the growth is likely to improve the reduced the defect density. Quantitatively, defect density is determined by text-book analysis of the low-temperature thermal conductivity in conjunction with the specific heat. The overall results indicate that the rubrene crystals are very clean with the defect density of only in the order of 10^{15} cm^{-3}. It appears that the cleanliness is partially responsible for their high performances when fabricated to single-crystal transistors.

ACKNOWLEDGMENTS

The authors thank T. Matsukawa, Y. Takahashi, T. Tokiyama, K. Sasai, Y. Murai, N. Hirotai, Y. Tominari, N. Mino, M. Yoshimura, M. Abe, Y. Kitaoka, Y. Mori, S. Morita, and T. Sasaki for the growth of the rubrene crystals grown from solutions. We also thank Y. Miyazaki for technical support in the specific heat measurement. This work was financially supported by a Grant-in-Aid for Scientific Research (Nos. 17069003 and 19360009) from MEXT, Japan.

REFERENCES

1. C. Kloc, P. G. Simpkins, T. Siegrist, and R. A. Laudise, J. Cryst. Growth **182**, 416 (1997).
2. V. Podzorov, V. M. Pudalov, and M. E. Gershenson, Appl. Phys. Lett. **82**, 1739 (2003).
3. J. Takeya, T. Nishikawa, T. Takenobu, S. Kobayashi, Y. Iwasa, T. Mitani, C. Goldmann, C. Krellner, and B. Batlogg, Appl. Phys. Lett. **85**, 5078 (2004).
4. T. Matsukawa, Y. Takahashi, T. Tokiyama, K. Sasai, Y. Murai, N. Hirotai, Y. Tominari, N. Mino, M. Yoshimura, M. Abe, J. Takeya, Y. Kitaoka, Y. Mori, S. Morita, and T. Sasaki, Jap. J. Appl. Phys. **47**, 8950 (2008).
5. J. Takeya, M. Yamagishi, Y. Tominari, R. Hirahara, Y. Nakazawa, T. Nishikawa, T. Kawase, T. Shimoda, and S. Ogawa, Appl. Phys. Lett. **90**, 102120 (2007).
6. R. Berman, *Thermal Conduction in Solids* (Oxford University Press, Oxford, 1976).

Mater. Res. Soc. Symp. Proc. Vol. 1154 © 2009 Materials Research Society 1154-B10-73

Increase in Open-Circuit Voltage and Improved Stability of Organic Solar Cells by Inserting a Molybdenum Trioxide Buffer Layer

Hideyuki Murata, Yoshiki Kinoshita, Yoshihiro Kanai, Toshinori Matsushima and Yuya Ishii
School of Materials Science, Japan Advanced Institute of Technology (JAIST)
1-1 Asahidai, Nomi, Ishikawa 923-129, Japan

ABSTRACT

We report an increase in open-circuit voltage (V_{oc}) by inserting an MoO_3 layer on ITO substrate to improve built-in potential of organic solar cells (OSCs). In the OSCs using 5,10,15,20-tetraphenylporphyrine (H_2TPP) as p-type material and C_{60} as n-type material, the V_{oc} effectively increased from 0.57 to 0.97 V with increasing MoO_3 thickness. The obtained highest V_{oc} (0.97 V) is consistent with the theoretical value estimated from the energy difference between the LUMO (-4.50 eV) of C_{60} and the HOMO (-5.50 eV) of H_2TPP layer. Importantly, the enhancement in the V_{oc} was achieved without affecting the short-circuit current density (J_{sc}) and the fill-factor (*FF*). Thus, the power conversion efficiency of the device increased linearly from 1.24% to 1.88%. We also demonstrated that a MoO_3 buffer layer enhances the stability of OSCs after photo-irradiation. We have investigated the stability of OSCs using H_2TPP and *N,N'*-di(1-naphthyl)-*N,N'*-diphenylbenzidine as p-type layer. Both devices with MoO_3 layer showed improved stability. These results clearly suggest that the interface between ITO and p-type layer affects device stability.

INTRODUCTION

In recent years, organic solar cells have attracted much attention as a new inexpensive renewable energy source. The increase in power conversion efficiency (η_P) and the improvement of stability of organic solar cells are key issues in the development of organic solar cells. The η_P of the solar cells depends on three device parameters, such as the open-circuit voltage (V_{oc}), the short-circuit current density (J_{sc}), and the fill factor (*FF*). Among those parameters, the J_{sc} can be improved by the use of bulk heterojunctions (e.g., the composite of p-type and n-type materials) as an active layer in both polymer [1] and small molecule-based solar cells. However, there is no enhancement effect on the V_{oc} by the use of bulk heterojunction structure. For the further improvement of η_P, it is essential to enhance V_{oc}, while keeping the corresponding J_{sc}. It has been shown that the V_{oc} depends on the energy difference between the lowest unoccupied molecular orbital (LUMO) of the electron acceptor material and the highest occupied molecular orbital (HOMO) of the electron donor material [2–4]. Recently, Mutolo et al. reported that V_{oc} of the solar cells increased in double heterojunction solar cells composed of boron subphthalocyanine chloride (HOMO level= -5.60 eV) and C_{60} (LUMO level= -4.50 eV) [5]. We have reported that V_{oc} incresed by inserting a thin layer of CuPc and Zn-phthalocyanine (ZnPc) with higher HOMO level (-5.1 eV) at the interface of pentacene (HOMO= -5.0 eV) and C_{60} [6].

In this study, we found that a modification of ITO surface by a high work function metal oxide (molybdenum trioxide MoO_3) is very effective in increasing V_{oc}. We demonstrate the systematic control of V_{oc} as a function of the film thickness of MoO_3 buffer layer in organic solar cells. The open-circuit voltage increased from 0.57 to 0.97 V as the thickness of MoO_3 film increased from 0 to 50 nm in the device structure of indium-tin-oxide ITO/ MoO_3 (x nm) / 5,10,15,20-tetraphenylporphine (H_2TPP, 10 nm) /C_{60} (40 nm)/bathocuproine (10 nm) /Ag (100 nm). The values between V_{oc} and the ionization potential of MoO_3 (x nm) on ITO exhibit a linear relationship, where the work function values change from 4.92 to 5.92 eV with increasing x from 0 to 50 nm.We also found that a MoO_3 buffer layer enhances the stability of organic solar cells under photo-irradiation. We have investigated OSCs with the structure of ITO/ MoO_3 (0 or 20 nm)/ p-type layer/ C_{60} (40 nm)/ Bathocuproine (BCP) (10 nm)/ Ag (100), where we use H_2TPP and *N,N'*-di(1-naphthyl)-*N,N'*-diphenylbenzidine (α-NPD) as p-type layer. Without MoO_3 layer, the devices showed a dramatic decrease in initial η_p under the same measurement conditions. However, both devices with MoO_3 layer showed excellent stability under the photo-irradiation. These results clearly indicate that the degradation occurs at the interface between ITO and p-type layer.

EXPERIMENTAL DETAILS

The device structure studied in this work is ITO/MoO_3 (0, 1, 5, 10, 20, and 50 nm) /p-type layer (10 nm) /C_{60} (40 nm) /bathocuproine (BCP) (10 nm) /Ag (100 nm). (Fig. 1 (a)), where H_2TPP and α-NPD are used as p-type layers, C_{60} is used as n-type layer, and BCP is used as an exciton blocking layer. The devices were fabricated in the following ways: Glass substrates coated with an ITO layer were cleaned using ultrasonication in acetone, followed by ultrasonication in detergent, pure water, and isopropanol. The substrates were treated by UV ozone for 30 min and then annealed at 150°C for 10 min in air. MoO_3 and organic layers were successively vacuum-deposited under a base pressure of 10^{-6} Torr on the cleaned ITO layer. To complete the OSC structures, an Ag layer was vacuum-deposited through a shadow mask to define the active area of the devices to be 4 mm^2.

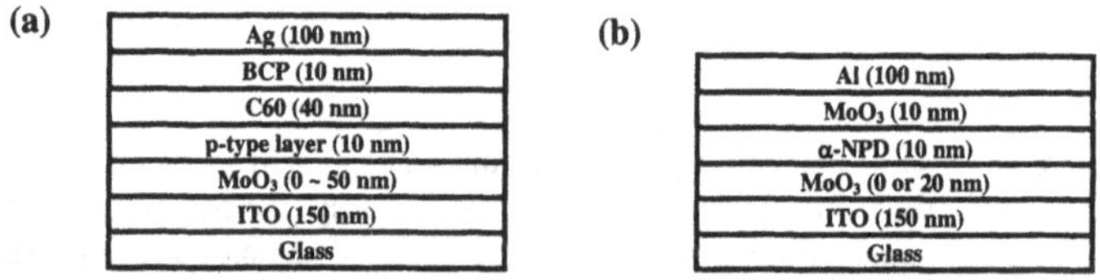

(a)

Ag (100 nm)
BCP (10 nm)
C60 (40 nm)
p-type layer (10 nm)
MoO_3 (0 ~ 50 nm)
ITO (150 nm)
Glass

(b)

Al (100 nm)
MoO_3 (10 nm)
α-NPD (10 nm)
MoO_3 (0 or 20 nm)
ITO (150 nm)
Glass

Figure 1. Schematics of structures of (a); organic solar cells and (b); hole-only devices.

To investigate the degradation mechanisms of the devices, the temporal change of the device characteristics after photo-irradiation was compared in the devices ITO/MoO_3 (0 and 20 nm) /p-type layer (10 nm)/C_{60} (40 nm)/BCP (10 nm)/Ag (100 nm).We also fabricated hole-only devices with a glass substrate/ITO (150 nm)/MoO_3 (0 or 20 nm)/α-NPD (70 nm)/MoO_3 (10 nm)/Al (100 nm) structure. In the hole-only structures, we used a high-work-function MoO_3 layer (-5.70 eV) at the α-NPD/Al interfaces to prevent injection of electrons from the cathode. The deposition rates were 0.03 nm/s for MoO_3, H_2TPP, α-NPD, and C_{60}, 0.1 nm/s for BCP, and 0.3 nm/s for Ag and Al. H_2TPP (99.0%) and C_{60} (99.5%) were purchased from Aldrich and MTR, Ltd., respectively. H_2TPP (Fig. 3) and C_{60} were sublimed in our laboratory before use. High purity BCP was provided by Nippon Steel Chemical Co., Ltd. and was used without further purification.

DISCUSSION

Dependence of V_{oc} on the thickness of MoO_3 layer

Figure 2a shows the current density-voltage (J-V) characteristics as a function of the film thickness of MoO_3. The V_{oc} drastically increased from 0.57 V to 0.97 V as the MoO_3 film thickness increased from 0 nm to 50 nm. The value of 0.97 V is close to the maximum value estimated from the energy difference between the LUMO (-4.50 eV) of C_{60} layer and the HOMO (-5.50 eV) of H_2TPP layer. The linear relationship between the work function of ITO/MoO_3 (x nm) and V_{oc} suggests that the observed increase in V_{oc} is the consequence of the enhancement of built-in potential generated between ITO/MoO_3 and Ag.

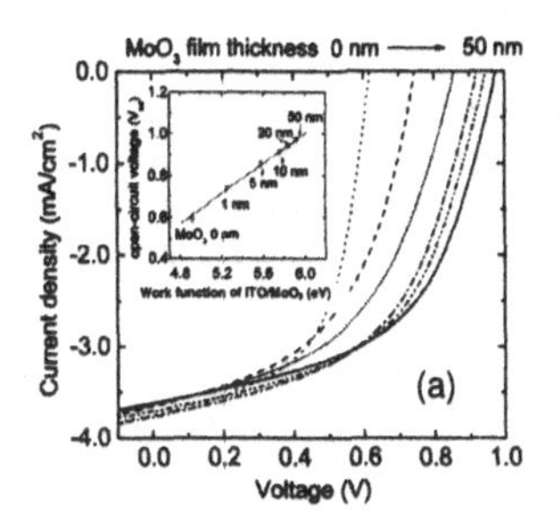

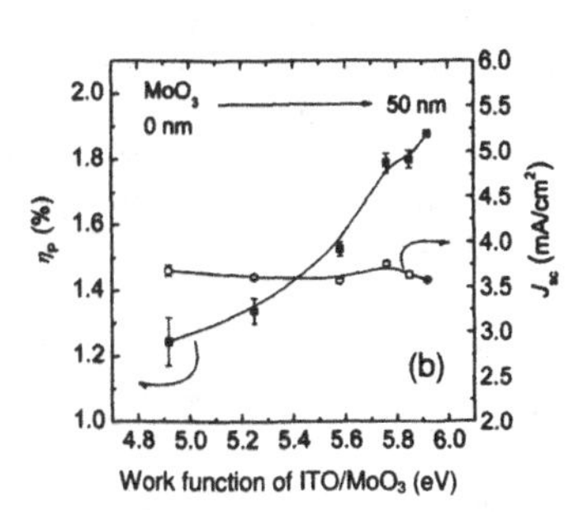

Figure 2. **(a)** Current density-voltage characteristics of ITO/H_2TPP (10 nm) /C_{60} (40 nm)/BCP (10 nm)/Ag (100 nm) under simulated AM1.5 solar illumination (100 mW/cm^2). Inset: V_{oc} versus work function of ITO/MoO_3 where the film thickness of MoO_3 was changed to 0, 1, 5, 10, 20, 50 nm. **(b)** The power conversion efficiency (η_p) and the short-circuit current density (J_{sc}) of the devices as a function of the work function of ITO/MoO_3.

Figure 2b shows the η_p and *Jsc* as a function of the work function of the ITO/MoO_3. The η_p of the device with MoO_3 layer increased from 1.24 % to 1.88% with increasing film thickness of MoO_3. On the other hand, the *Jsc* was independent of the film thickness of MoO_3. The *FF* of the device was also unchanged in the thickness range of 0 nm to 50 nm. (Data not shown) Thus, the enhancement of η_p is exclusively attributed to increase of V_{oc} by changing the film thickness of MoO_3. Since the enhancement of V_{oc} does not affect other device parameters, further improvement of η_p may be achieved in the devices with a bulk heterojunction interface.

According to a metal-insulator-metal model and a *p-n* junction model, the origin of V_{oc} can be explained in two ways: (1) the difference in energy level between work functions of an anode and a cathode and (2) the difference in energy level between a HOMO of a *p*-type material and a LUMO of a *n*-type material [2-4]. Since depositing the MoO_3 on the ITO surface increases the work functions of the anodes [7], the increase in the V_{oc} is attributable to an increase in work function of the ITO/MoO_3 anodes.

Effect of MoO_3 buffer layer on device stability

Figure 3 shows the changes of the J_{sc}, the V_{oc}, the FF, and the η_p for the OSCs with H_2TPP as a function of light irradiation time. While the OSCs with no buffer layer were drastically degraded with operational time, inserting the MoO_3 between the ITO and the H_2TPP suppressed the degradation. Although the η_p of the OSCs without the MoO_3 decreased to 37 % of its initial value after 60 min, the η_p of the OSC with the MoO_3 maintained 66 % of the initial value.

In addition to the H_2TPP OSCs, we investigated the stability of the α-NPD OSCs (Fig. 4). The η_p of the α-NPD OSC with the MoO_3 maintained 100 % of the initial value, while the α-NPD device without the MoO_3 decreased to 46 % of its initial value. In this device, we observed a slight increase in J_{sc}, FF, and η_p of the OSCs with MoO_3 after light irradiation. The origin of the increase in these parameters is not clear. As described later, we observed a similar increase of the dark current density in the hole-only α-NPD devices after light irradiation. One of the possible explanations of the increase in the J_{sc} might be a heating effect due to light irradiation [4]. These results clearly indicate that inserting the MoO_3 layer markedly improved OSC stability, and suggests that the degradation of the OSCs mainly occurs at the ITO/*p*-type layer interfaces (*Vide infra*).

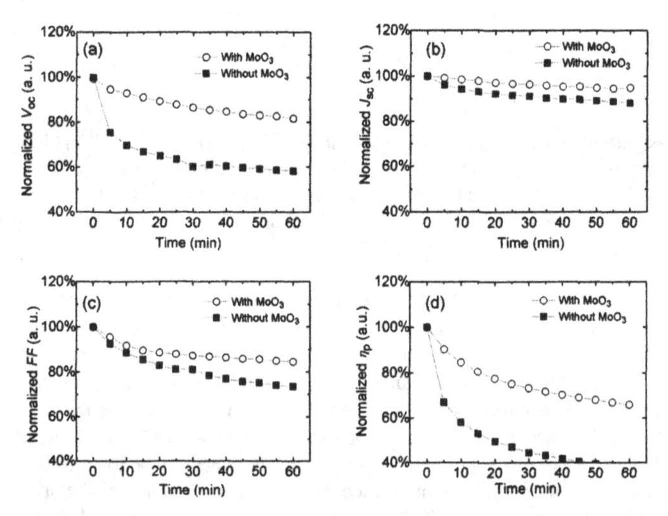

Figure 3. Changes of (a) V_{oc}, (b) J_{sc}, (c) FF, and (d) η_p for H_2TPP OSCs with and without MoO_3 buffer layer under light irradiation.

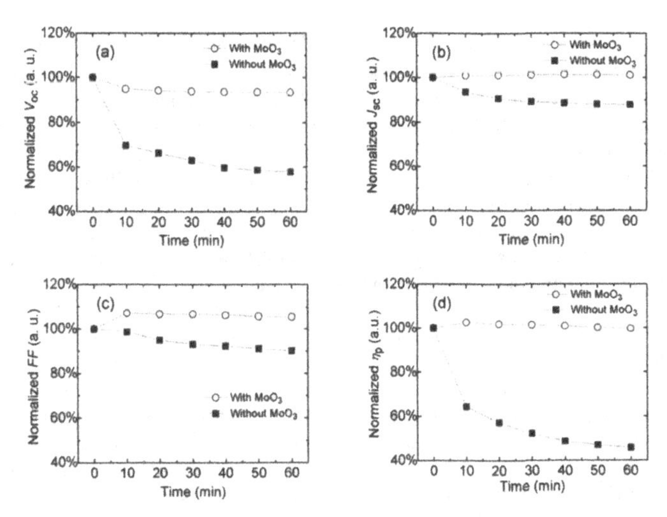

Figure 4. Changes of (a) V_{oc}, (b) J_{sc}, (c) *FF*, and (d) η_p for α-NPD OSCs with and without MoO_3 buffer layer under light irradiation.

To get insights about the anode/α-NPD interfacial degradation, we fabricated the hole-only α-NPD devices. Figure 5 shows the changes of the *J-V* characteristics of the hole-only devices during light irradiation. The hole-only α-NPD device without the MoO_3 layer was markedly degraded by light irradiation. The current density at the forward bias, where ITO electrode was biased positively, significantly dropped to 0.7% of its initial value, while the current density at the reverse bias was unchanged. On the other hand, we observed no degradation in the device with the MoO_3 layer, in either forward or reverse bias directions, after light irradiation. Instead, we observed a slight increase in current density due to light irradiation.

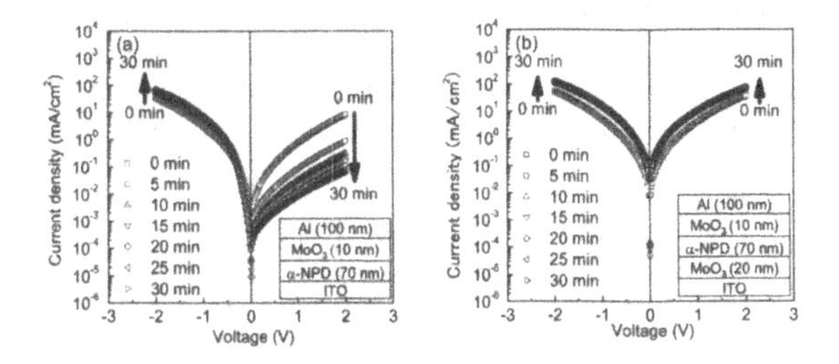

Figure 5. Change of *J-V* characteristics of α-NPD hole-only devices (a) without MoO_3 and (b) with MoO_3 under light irradiation. Insets show structures of hole-only devices.

These results suggest that the decrease in the current density in the forward bias region is not caused by an increase in resistance of the α-NPD bulk layer, but is caused by an increase in hole injection barrier height between the ITO and the α-NPD after light irradiation. In other words, the relative position of the HOMO level of the α-NPD to the Fermi level of the ITO might be shifted in the direction of increasing the hole injection barrier height after light irradiation. We attribute the shift of the relative energy level position to a vacuum level shift caused by a chemical reaction between the ITO and the α-NPD [8,9]. Moreover, the difference in Fermi levels of ITO and Ag are probably reduced by the vacuum level shift. In this case, the built-in potential of the OSCs decreases, resulting in the reduction of the V_{oc}. The insertion of the MoO_3 would prevent this reduction and, therefore, enhance device stability.

CONCLUSIONS

In conclusion, we have demonstrated that the V_{oc} increased from 0.57 to 0.97 V with the MoO_3 film thickness, due to enhanced built-in potential with increasing MoO_3 thickness. The highest V_{oc} (0.97 V) is consistent with the theoretical value estimated from the energy difference between the LUMO of C_{60} layer and the HOMO of H_2TPP layer. We have found that the OSC degradation occurs at the ITO and α-NPD interface under light irradiation, and that the degradation can be prevented by inserting a MoO_3 layer between ITO and the p-type layer. These findings would be beneficial for developing high performance organic solar cells.

REFERENCES

1. W. Ma, C. Yang, X. Gong, K. Lee and A. J. Heeger, *Adv. Funct. Mater.* **15**, 1617 (2005).
2. V. D. Mihailetchi, P. W. M. Blom, J. C. Hummelen, and M. T. Rispens, *J. Appl. Phys.* **94**, 6849 (2003).
3. C. J. Brabec, A. Cravino, D. Meissner, N. S. Sariciftci, T. Fromherz, M. T. Rispens, L. Sanchez, and J. C. Hummelen, *Adv. Funct. Mater.* **11**, 374 (2001).
4. C. J. Brabec, *Sol. Energy Mater. Sol. Cells* **83**, 273 (2004).
5. K. L. Mutolo, E. I. Mayo, B. P. Rand, S. R. Forrest, and M. E. Thompson, *J. Am. Chem. Soc.* **128**, 8108 (2006).
6. Y. Kinoshita, T. Hasobe, and H. Murata, *Appl. Phys. Lett.* **91**, 083518 (2007).
7. Y. Kinoshita, R. Tanaka and H. Murata, *Appl. Phys. Lett.*, **92**, 243309 (2008).
8. K. Akedo, A. Miura, K. Noda, H. Fujikawa, *Proceedings of the 13th International Display Workshops* (2006) 465.
9. H. Ishii, K. Sugiyama, E. Ito, K. Seki, *Adv. Mater.* **11,** 605 (1999).

Mater. Res. Soc. Symp. Proc. Vol. 1154 © 2009 Materials Research Society 1154-B10-92

Formation of Ohmic Carrier Injection at Anode/Organic Interfaces and Carrier Transport Mechanisms of Organic Thin Films

Toshinori Matsushima, Guang-He Jin, Yoshihiro Kanai, Tomoyuki Yokota, Seiki Kitada, Toshiyuki Kishi, and Hideyuki Murata *

School of Materials Science, Japan Advanced Institute of Science and Technology, 1-1 Asahidai, Nomi, Ishikawa 923-1292, Japan

*Corresponding author. Tel.: +81 761 51 1531; fax: +81 761 51 1149;
E-mail address: murata-h@jaist.ac.jp

ABSTRACT

We have shown that hole mobilities of a wide variety of organic thin films can be estimated using a steady-state space-charge-limited current (SCLC) technique due to formation of Ohmic hole injection by introducing a very thin hole-injection layer of molybdenum oxide (MoO_3) between an indium tin oxide anode layer and an organic hole-transport layer. Organic hole-transport materials used to estimate hole mobilities are 4,4',4"-tris(*N*-3-methylphenyl-*N*-phenyl-amino)triphenylamine (m-MTDATA), 4,4',4"-tris(*N*-2-naphthyl-*N*-phenyl-amino)triphenylamine (2-TNATA), rubrene, *N,N*′-di(m-tolyl)-*N,N*′-diphenylbenzidine (TPD), and *N,N*′-diphenyl-*N,N*′-bis(1-naphthyl)-1,1′-biphenyl-4,4′-diamine (α-NPD). These materials are found to have electric-field-dependent hole mobilities. While field dependence parameters (β) estimated from SCLCs are almost similar to those estimated using a widely used time-of-flight (TOF) technique, zero field SCLC mobilities (μ_0) are about one order of magnitude lower than zero field TOF mobilities.

INTRODUCTION

Organic light-emitting diodes (OLEDs) have been developed due to their high potentials for use in low-cost, mechanically flexible, light-weight display and lighting applications. Multilayer OLEDs are typically composed of an indium tin oxide (ITO) anode, an organic hole-transport layer (HTL), an emitting layer, an electron-transport layer, and a metal cathode. In general, a large hole injection barrier height of several hundred meV is present between an ITO layer and an HTL, which causes an increase in driving voltage of OLEDs. Various organic and inorganic hole-injection layers (HILs) have been inserted between an ITO and an HTL to reduce the driving voltages [1-3]. Recently, we have demonstrated that the use of a 0.75 nm HIL of molybdenum oxide (MoO_3) inserted between an ITO and an HTL of *N,N*′-diphenyl-*N,N*′-bis(1-naphthyl)-1,1′-biphenyl-4,4′-diamine (α-NPD) leads to the formation of an Ohmic contact at the ITO/MoO_3/α-NPD interfaces and the observation of a space-charge-limited current (SCLC) of α-NPD [4]. This MoO_3 thickness of 0.75 nm is much thinner than the previously reported values. Moreover, marked improvements of driving voltages, power conversion efficiencies, and operational stability of OLEDs have been realized using the very thin MoO_3 HIL [5,6].

Presence of charge-carrier injection barriers at electrode/organic interfaces generally makes it difficult to investigate carrier transport mechanisms in organic films because observed currents are governed by both carrier injection and transport [7]. In this study, we have shown that Ohmic contacts can be formed and SCLCs can be observed using a very thin MoO_3 HIL between an ITO anode layer and a wide variety of organic HTLs. Organic HTL materials used in this study are 4,4',4"-tris(*N*-3-methylphenyl-*N*-phenyl-amino)triphenylamine (m-MTDATA), 4,4',4"-tris(*N*-2-naphthyl-*N*-phenyl-amino)triphenylamine (2-TNATA), rubrene, *N*,*N*′-di(m-tolyl)-*N*,*N*′-diphenylbenzidine (TPD), α-NPD, 5,10,15,20-tetraphenylporphyrin (H_2TPP), and 2,4,6-tricarbazolo-1,3,5-triazine (TRZ-2). The chemical structures of the organic HTL molecules are shown in figure 1. From analyses of current density-voltage (*J*-*V*) characteristics of the HTLs with a SCLC equation, we found that while field dependence parameters (β) estimated from SCLC regions are almost similar to those estimated using a widely used time-of-flight (TOF) technique, zero field SCLC mobilities (μ_0) are about one order of magnitude lower than zero field TOF mobilities.

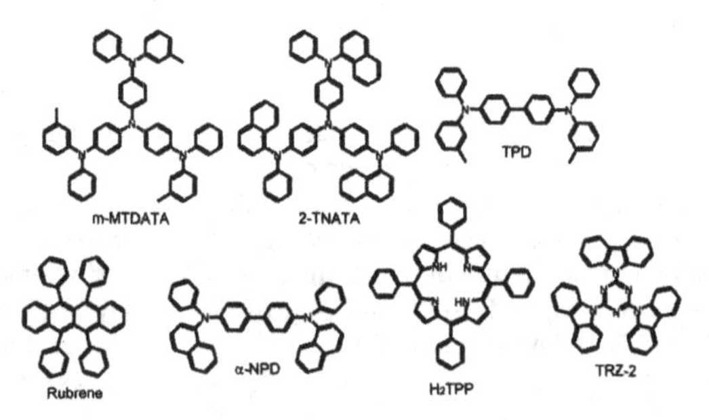

Figure 1. Chemical structures of organic molecules used to estimate hole mobilities.

EXPERIMENT

The schematic structure of the single-organic-layer, hole-carrier-only devices used in this study are shown in figure 2(a). The hole-only devices consisted of a glass substrate coated with a 150 nm ITO anode layer, a *X* nm MoO_3 HIL, a 100 nm organic HTL, an electron-blocking layer (EBL) of MoO_3, and a 100 nm cathode layer of Al. The devices were fabricated using a preparation condition as previously reported in Refs. [4-6]. The organic materials shown in figure 1 were used as the HTLs. The thicknesses of the MoO_3 HIL (*X*) between the ITO and the HTL were varied from 0 nm to 10 nm. The completed devices were encapsulated using a glass cap and an ultraviolet curing epoxy resin inside a nitrogen-filled glove box. The steady-state *J*-*V* characteristics of the devices with various *X* were measured using a SCS4200 semiconductor characterization system (Keithley) under dark at room temperature.

50 nm layers of the organic and inorganic materials were prepared on cleaned ITO surfaces to determine their ionization potential and work function energy levels by using an AC-2 photoelectron yield spectrometer (Riken Keiki). The electron affinity energy levels of the

organic layers were roughly estimated by subtracting their optical absorption onset energies from the ionization potential energy levels. The energy-level diagram of the devices with the values thus measured is depicted in figure 2(b). The hole injection and transport characteristics of the devices were investigated by changing the ionization potential energy levels of the HTLs from -5.06 eV (m-MTDATA) to -5.13 eV (2-TNATA), -5.24 eV (TPD), -5.29 eV (rubrene), -5.40 eV (α-NPD), -5.52 eV (H_2TPP), and -5.68 eV (TRZ-2) relative to the work function of ITO (-5.02 eV). As shown in figure 2(b), high-work-function MoO_3 (-5.68 eV) was introduced as the EBL between the HTL and the Al cathode to prevent electron injection from the cathode. In fact, all devices exhibited no electroluminescence from the HTLs during the *J-V* measurements, indicating that unipolar hole currents flow through the devices.

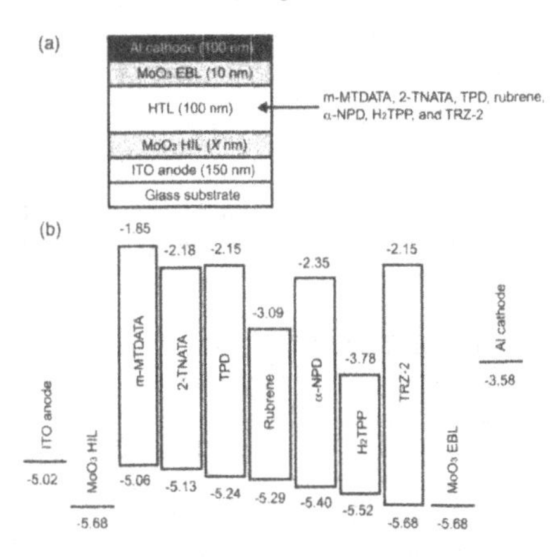

Figure 2. (a) Schematic structure of hole-only devices and (b) energy-level diagram of hole-only devices. Thicknesses of MoO_3 HIL used between ITO and HTL were varied from 0 nm to 10 nm.

RESULTS AND DISCUSSION

If an Ohmic contact is formed and a free charge density is negligible in comparison to an injected charge density, a SCLC with field-dependent carrier mobilities [8,9] is given by

$$J = \frac{9}{8}\varepsilon_r\varepsilon_0\mu_0 \exp(0.89\beta\left(\frac{V}{L}\right)^{0.5})\frac{V^2}{L^3} \quad (1),$$

where ε_r is the relative permittivity, ε_0 is the vacuum permittivity, μ_0 is the zero field mobility, β is the field dependence parameter, and L is the cathode-anode spacing. By fitting the *J-V* characteristics of the hole-only devices with Eq. (1), μ_0 and β values of the HTLs can be estimated at the same time. In this study, a standard ε_r of 3.0 for organic thin films was used to fitting.

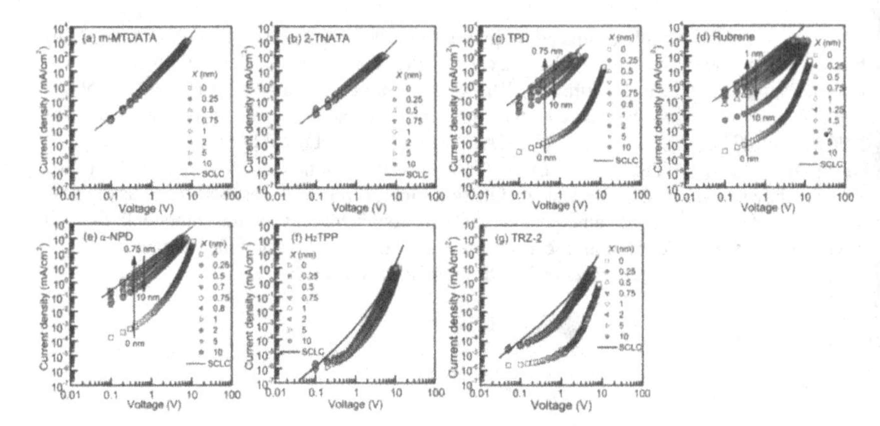

Figure 3. *J-V* characteristics of hole-only devices with HTLs of (a) m-MTDATA, (b) 2-TNATA, (c) TPD, (d) rubrene, (e) α-NPD, (f) H_2TPP, and (g) TRZ-2. The solid curves represent calculated *J-V* curves based on Eq. (1).

The *J-V* characteristics of m-MTDATA and 2-TNATA are independent of the oxide thicknesses *X* and are well fitted with Eq. (1) [the solid curves in figures 3(a) and 3(b)]. The μ_0 and β estimated by the fitting are (1.5 ± 0.1) x 10^{-6} cm^2 V^{-1} s^{-1} and (3.7 ± 0.4) x 10^{-3} $cm^{0.5}$ $V^{-0.5}$ for m-MTDATA and (4.5 ± 0.3) x 10^{-6} cm^2 V^{-1} s^{-1} and (2.0 ± 0.2) x 10^{-3} $cm^{0.5}$ $V^{-0.5}$ for 2-TNATA, respectively. These results indicate that the *J-V* characteristics are controlled by the SCLCs and the ITO/m-MTDATA and ITO/2-TNATA junctions are already Ohmic contacts without the MoO_3 HILs. The hole injection barrier heights are estimated from figure 2(b) to be 0.04 eV at the ITO/m-MTDATA interface and 0.11 eV at the ITO/2-TNATA interface. We assume that the very small barrier heights less than ≈ 0.1 eV no longer prevent hole injection from the ITO. It has been reported that ITO forms ideal hole-injecting contacts with m-MTDATA [10] and 2-TNATA [11], which is in good agreement with the present results.

The *J-V* characteristics of TPD, rubrene, and α-NPD are markedly dependent upon the *X*. The *J-V* characteristics of these HTLs shift to higher current densities and then shift to lower current densities as increasing the *X* from 0 nm to 10 nm [figures 3(c), 3(d), and 3(e)]. The similar change in *J-V* characteristics has been observed [4]. The optimized *X* is 1.0 nm for rubrene and 0.75 nm for TPD and α-NPD, which provide the highest current densities for the devices. The *J-V* characteristics at the optimized *X* are well fitted with Eq. (1) [the solid curves in figures 3(c), 3(d), and 3(e)], yielding a μ_0 of (4.5 ± 0.4) x 10^{-5} cm^2 V^{-1} s^{-1} and a β of (1.5 ± 0.2) x 10^{-3} $cm^{0.5}$ $V^{-0.5}$ for TPD, a μ_0 of (1.0 ± 0.1) x 10^{-4} cm^2 V^{-1} s^{-1} and a β of (1.1 ± 0.2) x 10^{-3} $cm^{0.5}$ $V^{-0.5}$ for rubrene, and a μ_0 of (8.1 ± 0.6) x 10^{-5} cm^2 V^{-1} s^{-1} and a β of (8.0 ± 0.6) x 10^{-4} $cm^{0.5}$ $V^{-0.5}$ for α-NPD.

As can been seen in figure 2(b), the relatively large hole injection barriers are present between the ITO and the HTLs of TPD, rubrene, and α-NPD. However, these barrier heights are lowered by using the very thin MoO_3 HIL to form Ohmic hole injection due probably to an increase in work function of an ITO/MoO_3 anode [6] and efficient hole injection via gap states caused by an electron transfer from the HTLs to the MoO_3 HIL [12]. On the other hands, we suppose that a strong space charge layer is formed in the HTLs due to the electron transfer when the thick MoO_3 HIL is used. The strong space charge layer may lower carrier injection at the interfaces, resulting in a decrease in current density at higher X, as shown in figures 3(c), 3(d), and 3(e).

The J-V characteristics of the hole-only devices with the H_2TPP HTL and the TRZ-2 HTL are shown in figures 3(f) and 3(g). The J-V characteristics are not well explained with Eq. (1) [the solid lines in figures 3(f) and 3(g)], indicating that the J-V characteristics are controlled by injection-limited currents. Since the ionization potential energies of H_2TPP and TRZ-2 are deeper than those of the other HTL materials (figure 2), Ohmic conductions are not realized even by using the MoO_3 HIL.

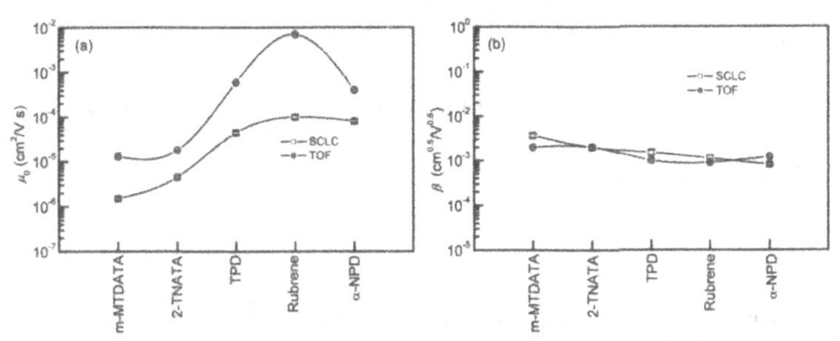

Figure 4. Comparison of μ_0 and β estimated from SCLC and TOF techniques.

The μ_0 and β estimated from the SCLC regions are compared with those of m-MTDATA [13], 2-TNATA [11], TPD [14], rubrene [15], and α-NPD [14] previously estimated using a TOF technique. The comparison results of the μ_0 and the β are shown in figures. 4(a) and 4(b), respectively. While the β estimated from the SCLCs is almost similar to that estimated using a TOF technique, the μ_0 estimated from the SCLCs is about one order of magnitude lower than that estimated using a TOF technique. It has been shown that electron mobilities of tris(8-hydroxyquinoline) aluminum (Alq_3) [16] and 4,7-diphenyl-1,10-phenanthroline (Bphen) [17] and hole mobilities of α-NPD [18] and m-MTDATA [19] gradually decrease with decreasing their film thicknesses. Chu *et al.* attributed the thickness-dependent mobilities to a change of hole trap concentrations in films [18]. Thus, we infer that SCLC mobilities measured from thinner films become lower than TOF mobilities measured from thicker films.

CONCLUSIONS

In this study, we investigated that how thicknesses of a MoO_3 HIL inserted between an ITO layer and a wide variety of organic HTLs influence J-V characteristics of hole-only devices. We obtained the following relationships between hole injection barrier heights and J-V characteristics. (1) Small barrier heights less than ≈ 0.1 eV at interfaces of ITO/m-MTDATA and ITO/2-TNATA no longer prevent hole injection from the ITO (Ohmic contacts). Thus, SCLCs of m-MTDATA and 2-TNATA are observed without a MoO_3 HIL. (2) Barrier heights ranging from ≈ 0.1 eV to ≈ 0.5 eV at interfaces of ITO/TPD, ITO/rubrene, and ITO/α-NPD lower hole injection. However, introduction of a very thin MoO_3 HIL at the interfaces leads to formation of Ohmic contacts and observation of their SCLCs. (3) Barrier heights larger than ≈ 0.5 eV at interfaces of ITO/H_2TPP and ITO/TRZ-2 markedly lower hole injection. Ohmic contacts are not realized at the interfaces even by using a MoO_3 HIL. Moreover, by analyzing *J*-*V* characteristics of the organic HTLs with a SCLC equation, we obtained μ_0 and β, as summarized in table I. The μ_0 estimated from the SCLC regions is found to be lower than those estimated using a TOF technique. We believe that these findings are indispensable to clarifying carrier injection and transport mechanisms of organic thin films and to developing OLEDs.

Table I. μ_0 and β estimated from SCLCs

HTL material	μ_0 ($cm^2 V^{-2} s^{-1}$)	β ($cm^{0.5} V^{-0.5}$)
m-MTDATA	$(1.5 \pm 0.1) \times 10^{-6}$	$(3.7 \pm 0.4) \times 10^{-3}$
2-TNATA	$(4.5 \pm 0.3) \times 10^{-6}$	$(2.0 \pm 0.2) \times 10^{-3}$
TPD	$(4.5 \pm 0.4) \times 10^{-5}$	$(1.5 \pm 0.2) \times 10^{-3}$
Rubrene	$(1.0 \pm 0.1) \times 10^{-4}$	$(1.1 \pm 0.2) \times 10^{-3}$
α-NPD	$(8.1 \pm 0.6) \times 10^{-5}$	$(8.0 \pm 0.6) \times 10^{-4}$

REFERENCES

(1) S. A. VanSlyke, C. H. Chen, and C. W. Tang, *Appl. Phys. Lett.* **69**, 2160 (1996).
(2) S.-F. Chen and C.-W. Wang, *Appl. Phys. Lett.* **85**, 765 (2004).
(3) S. Tokito, K. Noda, and Y. Taga, *J. Phys. D: Appl. Phys.* **29**, 2750 (1996).
(4) T. Matsushima, Y. Kinoshita, and H. Murata, *Appl. Phys. Lett.* **91**, 253504 (2007).
(5) T. Matsushima and H. Murata, *J. Appl. Phys.* **104**, 034507 (2008).
(6) T. Matsushima, G.-H. Jin, and H. Murata, *J. Appl. Phys.* **104**, 054501 (2008).
(7) M. Abkowitz, J. S. Facci, and J. Rehm, *J. Appl. Phys.* **83**, 2670 (1998).
(8) M. A. Lampert and P. Mark, *Current Injection In Solids* (ACADEMIC, New York, 1970).
(9) P. N. Murgatroyd, *J. Phys. D: Appl. Phys.* **3**, 151 (1970)
(10) C. Giebeler, H. Antoniadis, D. D. C. Bradley, and Y. Shirota, *Appl. Phys. Lett.* **72**, 2448 (1998).
(11) C. H. Cheung, K. C. Kwok, S. C. Tse, and S. K. So, *J. Appl. Phys.* **103**, 093705 (2008).
(12) H. Lee, S. W. Cho, K. Han, P. E. Jeon, C.-N. Whang, K. Jeong, K. Cho, and Y. Yi, *Appl. Phys. Lett.* **93**, 043308 (2008).
(13) S. W. Tsang, S. K. So, and J. B. Xu, *J. Appl. Phys.* **99**, 013706 (2006).
(14) S. Naka, H. Okada, H. Onnagawa, Y. Yamaguchi, and T. Tsutsui, *Synth. Met.* **111-112**, 331 (2000).
(15) H. H. Fong, S. K. So, W. Y. Sham, C. F. Lo, Y. S. Wu, and C. H. Chen, *Chem. Phys. Lett.* **298**, 119 (2004).
(16) S. C. Tse, H. H. Fong, and S. K. So, *J. Appl. Phys.* **94,** 2033 (2003).
(17) W. Xu, Khizar-ul-Haq, Y. Bai, X. Y. Jiang, and Z. L. Zhang, *Solid State Communications* **146,** 311 (2008).
(18) T. Y. Chu and O. K. Song, *Appl. Phys. Lett.* **90**, 203512 (2007).
(19) O. J. Weiß, R. K. Krause, and A. Hunze, *J. Appl. Phys.* **103,** 043709 (2008).

Mater. Res. Soc. Symp. Proc. Vol. 1154 © 2009 Materials Research Society 1154-B10-95

Accurate and Simultaneous Determination of Carrier Density and Mobility in Organic Semiconducting Materials

Kai Shum[1, (a)], Zhuo Chen[1], C. M. Xue[2], and Shi Jin[2, (b)]

[1]Department of Physics, Brooklyn College of CUNY, Brooklyn, NY 11020, U.S.A.

[2]Department of Chemistry, College of Staten Island, CUNY, Staten Island, NY 10314, U.S.A.

(a) kshum@brooklyn.cuny.edu
(b) jin@csi.cuny.edu

ABSTRACT

How to accurately determine carrier mobility and density in organic semiconducting materials is a very important subject for their optoelectronic applications including light-emitting diodes, solar cells, and thin film field-effect transistors. In this work, we report on a unique data analysis procedure for space-charge limited currents to simultaneously obtain the carrier density and mobility in semiconducting organic-materials. This procedure has been used for a few newly synthesized perylene tetracarboxylic diimide (**PDI**) derivatives with tunable π-stack structures without altering the electronic characteristic of individual molecules. How π-stack structural variation and residual carrier density affect electron transport performance will be discussed.

INTRODUCTION

Carrier mobility and density in organic materials are very important parameters for various electronic applications including organic light-emitting diodes (OLEDs), solar cells, and thin film field-effect transistors. Therefore, accurate determinations of these two parameters in unintentionally doped pristine organic materials are crucial to understand the carrier transport physics and to fully control the chemical synthesis in terms of molecular designs.

There are two main methods to determine the carrier mobility in organic materials: 1) time-of-flight method, [1] and 2) steady-state space-charge-limited current (SCLC) method. [2] In the first method, both electrons and holes are generated by short light pulses in the organic materials sandwiched by two conducting electrodes. For electron (hole) mobility measurements, the light excitation is set to be near the negative (positive) electrode. Transit time, Δt_{ransit} for either electrons or holes optically generated near one electrode drifted across the device to another electrode with the thickness of L is then measured from time-resolved photocurrent data. Carrier mobility, μ_c, is then determined by the following simple equation:

$$\mu_c E = L/\Delta t_{ransit}. \qquad (1)$$

The electric field, E, in the above equation is related the applied voltage, V, by E = V/L. This method assumes uniform electric field within the device. Its validation depends on the density level of photogenerated carriers.

In the second method, the conventional steady-state current is measured as a function of applied bias on the same device as in the first method except that one of electrodes does not have to be transparent necessary for photo-excitation. For small applied voltages, the measured current follows Ohm's law: J = σE, where σ is related carrier charge, q, carrier density, n_c, and carrier mobility by $\sigma = qn_c\mu_c$. Electrodes act to hold electric charges provided by an external source to maintain applied voltage across the device and to provide current continuity. In this applied voltage regime, carrier mobility can be obtained only when carrier density can be independently determined, which is usually difficult to do especially for newly synthesized materials. Above a certain critical applied voltage, the measured current density due to the injected carriers from an electrode (electrons from negative electrode while holes from positive electrode) may become *dominant* and quadratically dependent on the applied bias. Under this condition, a theoretical expression was initially obtained by Mott and Gurney [3] to describe the current density for a material sandwiched between two parallel plane electrodes:

$$J = \frac{9}{8}\epsilon_r\epsilon_0\mu_c\frac{V^2}{L^3}. \qquad (2)$$

In the Eq. (2), the carrier density in the material, n_c, can be considered being replaced by an equivalent injected charge density, n_{inj} (V) = (9/8) $\varepsilon_r\varepsilon_0V/(qL^2)$, in comparison with a conventional Ohm's law (J = σ E), where ϵ_r is the relative dielectric constant of the material and ϵ_0 is the permittivity of free space. Various extensions of this expression have been developed since the initial work in Ref. 3. When the carrier traps with the density of states (N_t) located at the energy, ΔE, below the lowest unoccupied molecular states (LUMO) are considered, the injected carrier density is reduced by a factor of θ. Hence, the Eq. (2) is modified as

$$J = \frac{9}{8}\theta\epsilon_r\epsilon_0\mu_c\frac{V^2}{L^3}. \qquad (3)$$

With the assumption Boltzmann carrier distribution, the carrier reduction factor θ is given by

$$\theta = \frac{1}{1+\frac{N_t}{N_c}e^{\Delta E/KT}}. \qquad (4)$$

Furthermore, when the reduction of trap potential is considered under a strong electric field (Frenkel Effect), an approximation for the space-charge limited current was obtained by PNM: [4]

$$J = \frac{9}{8}\theta\epsilon_r\epsilon_0\mu_c\exp\left(0.89\gamma\sqrt{V/L}\right)\frac{V^2}{L^3}. \qquad (5)$$

In the Eq. (5), the temperature dependent γ is given by $\left(\frac{1}{KT}\right)\sqrt{e^3/(\pi\epsilon_r\epsilon_0)}$. At room temperature, the value of γ is 0.00168 $V^{-0.5}m^{0.5}$ for $\epsilon_r = 3$. This Frenkel Effect becomes important only when the device is under a high bias resulting in an internal electric field on an order of 10^6 V/m. For an example, at the field of 4×10^6V/m, the value for $\exp\left(0.89\gamma\sqrt{\frac{V}{L}}\right) = 20$. It is important to note that the carrier reduction factor due to traps reduces the SCLC while the Frenkel effect enhances it resulting in a very complicated balance.

In the past, many researchers [5, 6] have used Eq. 5 to extract carrier mobility in various semiconducting organic materials. There are two drawbacks for using this method to obtain carrier mobility. First, in order to get quadratic V dependence, often, it requires fairly large applied voltage, more than usually used for device operation. In this large applied field range, the complicated Freckle effect may substantially reduce the accuracy of carrier mobility determination. Secondly, this method does not provide information on carrier density that is often as important as carrier mobility.

THEORETICAL APPROACH

In this paper, we describe a unique data analysis procedure to obtain simultaneously carrier density and mobility in an intermediate applied voltage range. In this procedure, we set $\theta = 1$ as other groups [5, 6] have done for the first order approximation. With this approximation, the measured current density consists of two major parts: ohmic current density due to the contribution of carriers in organic materials and the space-charge current (SCC) density due to injected carriers from an electrode. It can be expressed as:

$$J = q\mu_e (V/L) [n_e + (9/8)\, \varepsilon_r\varepsilon_0 (V/qL^2)]. \qquad (6)$$

The first term in the [ ] in the above equation is the material carrier density (from here on, we use subscript e to represents electron since we will deal with electron transport materials in this work) in a given organic material while the second term can be considered as the injected electron density from the cathode. From the Eq. (6), we can define a *cross-over voltage*, V_{co}, at which the current density due to the electrons in organic materials is equal to the injected charge density. This voltage is then given by

$$V_{co} = (8/9) [(qn_eL^2)/ (\varepsilon_r\varepsilon_0)]. \qquad (7)$$

From a log-log plot of experimentally measured J-V curve, we can initially estimate a value for V_{co} as shown in Fig. 2. With this initial value of V_{co}, the electron density can be calculated by Eq. (7). After obtaining the value of n_e, only parameter left in the Eq. (6) is the carrier mobility provided that the relative dielectric constant has been independently determined. The fitting can be repeated convergently by slightly changing the initial estimate of the cross-over voltage. Using this procedure, not only electron mobility can be accurately determined, but also electron density in semiconducting organic materials.

EXPERIMENTAL DETAILS

To validate our data fitting procedure, a few low molecular weight perylene tetracarboxylic diimide **(PDI)** derivatives were synthesized. There are two reasons for this choice. First, this molecule series are known to be good electron transporting materials with electron mobility of up to 2.1 $cm^2\ V^{-1}\ s^{-1}$. [7, 8] Second, inter-molecular separation, $S_{inter\text{-}m}$, could be finely tuned by changing different side chains. This inter-molecular separation may correlate with electron mobility. The Table 1 lists the four **PDI** derivatives we used. In addition, this procedure has also been used to fit the experimental data recently reported in the literature to get the electron mobility values for their materials. The PDI derivatives were synthesized according to a slightly modified literature procedure [9] with benzyl bromide replacing dodecyl bromide.

The devices were fabricated by sandwiching organic materials in their molten state between two conducting indium tin oxide (ITO) coated on glass slides in a nitrogen purged glove box. The thickness of devices was controlled by 10 micrometer glass spacers. JV data were measured by HH/Agilent 4142B modular DC source/monitor. Dielectric constants were determined by frequency dependence of impedance measurements using Princeton Applied Research 2273 potentiostat/galvanostat over a frequency range from100 Hz to 100 kHz. The separation distances between two adjacent molecules along the stacking axis direction were obtained from wide-angle X-ray scattering data.

Table 1 PDI derivatives

Abbreviation	**Full name**	**$S_{inter-m}$ (nm)**
DB-IsoLeu	*N, N'*-di((*S*)-2-phenylmethyloxyl-1-(1-methylpropyl)-2-oxoethyl)-3,4:9,10-perylenetetracarboxyldiimide	0.362
DB-Val	*N, N'*-di((*S*)-2-phenylmethyloxyl-1-(1-methylethyl)-2-oxoethyl)-3,4:9,10-perylenetetracarboxyldiimide	0.365
DB-Ala	*N, N'*-di((*S*)-2-phenylmethyloxyl-1-methyl-2-oxoethyl)-3,4:9,10-perylenetetracarboxyldiimide	0.347
DB-Leu	*N, N'*-di((*S*)-2-phenylmethyloxyl-1-(2-methylpropyl)-2-oxoethyl)-3,4:9,10-perylenetetracarboxyldiimide	0.346

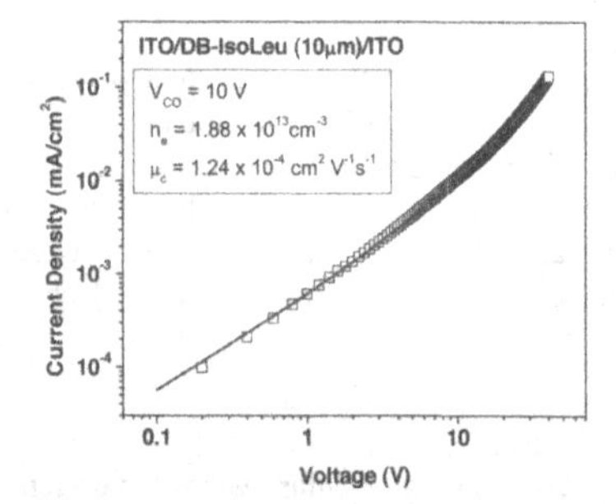

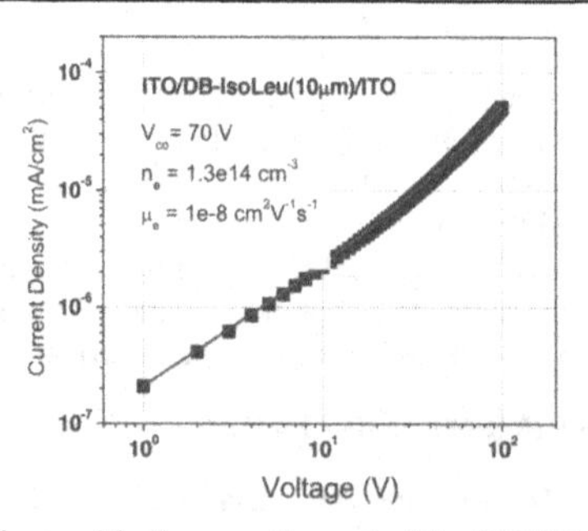

Fig. 1 Measured JV and the best fits for 2 devices with the sample material of DB-IsoLeu synthesized at 2 different times.

RESULTS AND DISCUSIONS

In the Fig. 1 to 3, the experimental data are shown as black discrete symbols and the calculated J-V curves are shown as solid red curves. Three parameters are shown in each figure, the value for V_{co}, n_e calculated using the Eq. (7), and finally, the carrier (electron) mobility is obtained by the best fit of the calculated J-V using the Eq. (6) to the measured J-V curves.

In Fig. 1, the data displayed in the left and right figures were taken from the same materials but synthesized at two different times. Difference values obtained for both electron density and mobility basically reflects the fact that material is not yet well controlled, especially impurity sites that produce residual electrons. Fig. 2 shows the measured JV curves from two different side chain materials synthesized at the same time: DB-Val (left) and DB-Ala (right). From their best fits, they have same order of magnitude for electron mobility.

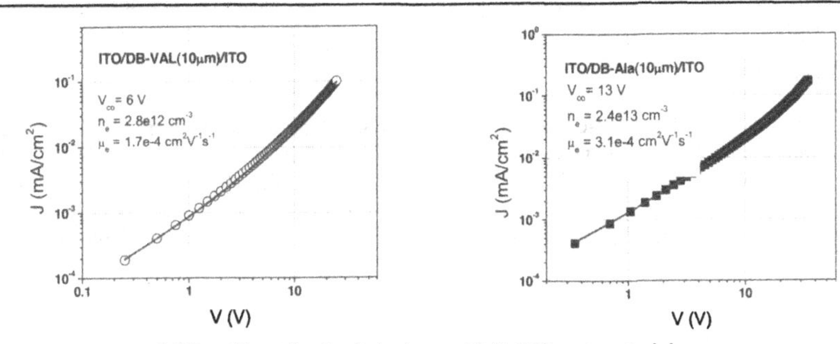

Fig. 2 Measured JV and best fits for 2 devices with 2 different materials.

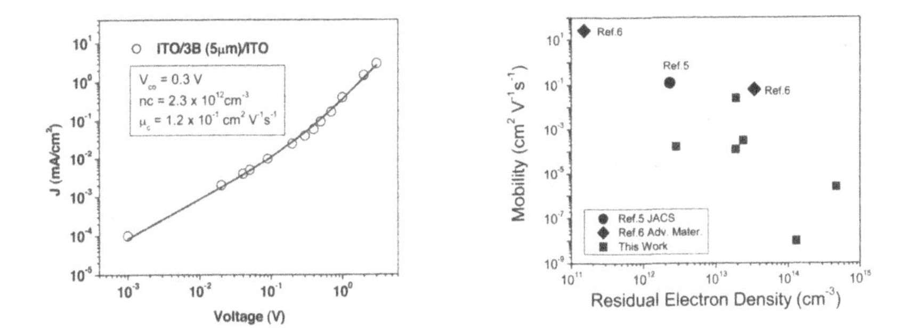

Fig. 3 Left: Measured JV and best fits for the devices given in Ref.5; Right: A summary of electron mobility for various devices plotted as a function of electron density.

Fig. 3 (left) displays the measured JV from a different material reported by other group. [5] An amorphous hexaazatrinaphthylene (HATNA) derivative was used for their device. By applying our fitting procedure, not only electron mobility of 0.12 cm^{-2} $V^{-1}s^{-1}$, but also residual electron density of 2.3 x 10^{12} cm^{-3}, was obtained without involving complicated Frenkel effect.

In Fig. 3 (right), a summary of electron mobility values are given and are plotted as a function of residual electron density.

Correction of electron mobility with inter-molecular spacing, $S_{inter-m}$, was attempted. However, due to tight range for the inter-molecular spacing achieved, it is not possible to have a clear correlation with measured electron mobility. From the deduced values of electron mobility for various devices plotted as a function of electron density, it is clear that the electron mobility decreases logarithmically as electron density in increases. This behavior may be consistent with a picture that electron mobility is limited by the number of impurity sites. These scattering sites also produce residual electrons, hence, higher residual electron density is, and lower electron mobility will be.

CONCLUSIONS

In summary, we have present a unique data analysis method in which carrier density and mobility can be obtained simultaneously. The method was applied to a few newly synthesized PDT electron transport derivatives. It was found the electron mobility decreases logarithmically with residual electron density. However, we failed to correlate electron mobility to inter-molecular spacing of pi-stacking molecules due to the limited tuning range we were able obtain.

ACKNOWLEDGMENTS

The authors would like to acknowledge CUNY-RF collaborative grant for this work.

REFERENCES

1. P. M. Borsenberger, D. S. Weiss, *Organic Photoreceptors for Xerography*; Marcel Dekker, New York, 1998.
2. Rose, A. Phys. Rev. 97, 1538 (1955).
3. N. F. Mott, and D. Gurney, *Electronic Processes in Ionic Crystals*, Academic Press, New York, 1970.
4. P N. J. Murgatroyd, Phys. D3, 151 (1970).
5. R. K. Bilal et al. J. AM. CHEM. SOC. 127, 16358 (2005).
6. Zesheng An et al., Adv. Mater. 17, 2580 (2005)
7. S. Tatemichi et al., Appl Phys. Lett. 89, 112108 (2006).
8. J. H. Oh et al. Appl Phys. Lett. 91, 212107 (2007).
9. Y. J. Xu, S. W. Leng, C. M. Xue, J. Pan, J. Ford and. S. Jin. Angew. Chem. Int. Ed. 46, 3896 (2007)

Mater. Res. Soc. Symp. Proc. Vol. 1154 © 2009 Materials Research Society 1154-B10-108

Towards Greatly Improved Efficiency of Polymer Light Emitting Diodes

Zhang-Lin Zhou, Xia Sheng, Lihua Zhao, Gary Gibson, Sity Lam, K. Nauka and James Brug*
Hewlett Packard Labs, 1501 Page Mill Road
Palo Alto CA 94304, U.S.A.

ABSTRACT

Polymer light-emitting diodes (PLEDs) show great promise of revolutionizing display technologies. The archetypical multilayer PLED heterostructure introduces numerous chemical and physical challenges to the develoment of efficient and robust devices. These layered structure are formed from solution based spin-casting or printing with subsequesnt removal of the solvent. However, solvent from the freshly deposited film may dissolve or partially dissolve the underlying layer resulting in loss of the desired structure and corresponding device functionality. Undesirable changes in the morphology and interfaces of the polymer films are another detrimental effect associated with solvent removal. Herein, we demonstrated that by embedding hole transporting materials (HTLs) in a cross-linked polymer matrix, the total luminance and external quantum efficiency were greatly improved over devices without this HTL layer.

INTRODUCTION

The growth and proliferation of electronic devices has created a significant industry-wide demand for new, low-power, light and low-cost display technologies. This demand underlies a current display development initiative within display industries. In the last decade, a number of intensive studies have been made to achieve efficient polymer light-emitting diodes (PLEDs). Chemistry and chemical principles have played a crucial role in the evolution of efficient PLEDs. As a result of extensive multidisciplinary efforts, modern PLEDs offer substantial benefits over conventional cathode ray tubes (CRT) and liquid crystal display (LCD). PLEDs display provides superior brightness and color purity, markedly lower power consumption, as well as full viewing angle without compromising image quality. Compared with small molecule organic LEDs (OLEDs), PLEDs use solution-based processes, which offer the potential for lower cost and roll-to roll processing on flexible substrates. To realize these favorable advantages, significant chemical and physiochemical challenges must be addressed. These challenges include (i) excellent solution-processable multilayer structures, (ii) improved efficiency via balanced charge carrier injection and leakage current reduction, (iii) better thermal stability and (iv) increased operational lifetime [1].

An efficient PLED device typically consists of a stack of organic/polymeric thin layers, each one of them performing a specific function aimed at improving the device performance or achieving the desired device functionality. In many cases, these layered structures are formed from the polymer solution by spin-casting or printing with subsequent removal of the solvent. However, solvent from the freshly deposited film frequently dissolve or partially dissolve the underlying layer, resulting in loss of the desired structure and corresponding device functionality. Undesirable changes in the morphology and interfaces of the polymer films are another detrimental effect associated with solvent removal. To make more robust hole transport layers

(HTLs) and avoid solvent damage from subsequent emissive layer, the most common approach is to introduce polymerizable functional groups onto the basic structure of the molecules with hole transporting (HT) property to form a cross-linkable HT molecules, which can form a cross-linked HTL upon spin coating [2-8]. However, this type of polymerizable hole transporting material is expensive and difficult to make. Herein we report a new approach to address this issue: Commercially available HT polymers are embedded into a cross-linked polymer network to "lock" uniformly distributed HT polymers inside the cross-linked polymer matrix. This approach proved to be more advantageous in terms of process simplicity and cost.

RESULTS AND DISCUSSION

The basic multilayer heterostructure and energy level diagram of PLED is shown in **Figure 1**. Components include a transparent conducting anode, hole injection layer (HIL), HTL, emissive layer (EML), electron-transporting layer (ETL), and metallic low work function cathode. For a convenient demonstration, we have chosen indium tin oxide (ITO) as anode, poly (3,4-ethlyenedixoythiophene)-poly(styrene sulfonic acid (PEDOT-PSS) as hole injection layer, poly(9,9- dioctylfluorene-co-N-(4-butylphenyl)diphenylamine (ADS132GE) as HTL, poly(9.9-dioctylfluorenyl-2,7-diyl) (PFO) as EML and Ba-Al as the cathode (chemical structures shown in **Figure 2**). This set of materials might not be optimum in terms of interfacial energy alignment for the transport of holes and electrons, but suffices for the demonstration of the function of our embedded HTL materials.

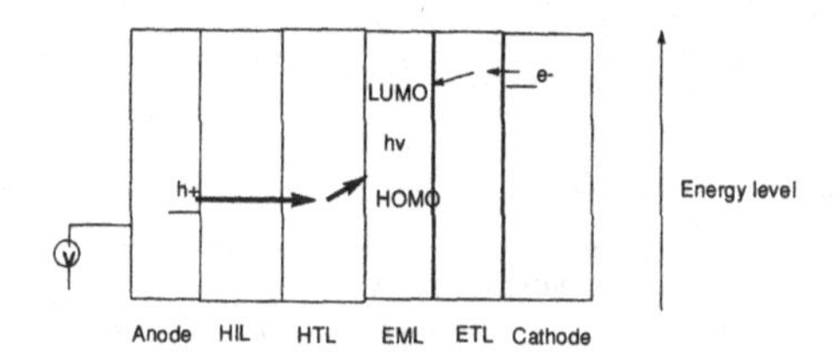

Figure 1 Schematic of typical PLED heterostrucutre

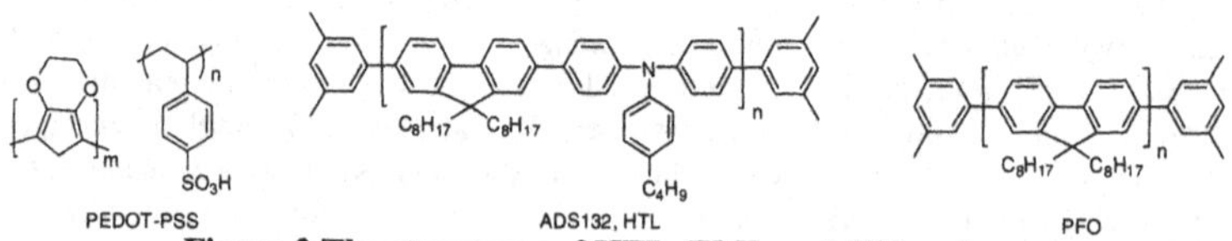

PEDOT-PSS ADS132, HTL PFO

Figure 2 The structures of HTL, EML and HIL polymers

In a demonstration device, the HIL, HTL, and EML layers were sequentially spun-cast onto a glass substrate with a pre-patterned ITO anode. A Ba-Al film was then thermally evaporated to form a cathode. As we know, solvents that are used for EML polymers are commonly shared by the under-layer HTL polymer. To minimize the undesirable impact from solvents that are used by the EML polymer, we embed the HTL molecules into the inert cross-linked polymer network. The cross-linkable polymer is selected so that the electrical and electro-optical properties of the embedded polymer are retained. At the same time, the cross-linked polymer network helps maintain the morphology of the embedded organic film during solvent removal. There are many options could be employed to form the cross-linked inert polymer network: a mixture of cross-linkable monomer, oligomers, and polymers, in addition to cross-linking agent and an initiator. The cross-linking agent could be a 2-branch, 3-branch, or 4-branch cross-linker. Cross-linking could be activated using appropriate energy sources such as thermal process or UV-exposure. Some examples of the cross-linkers are shown in **Figure 3**.

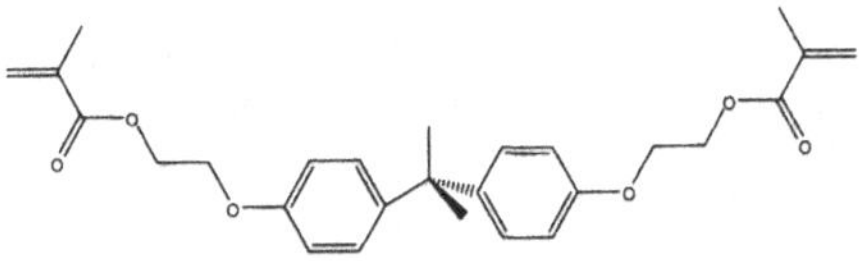

Ethoxylated (2) Bisphenol A Dimethylacrylate, EBDA
2-branch cross linker

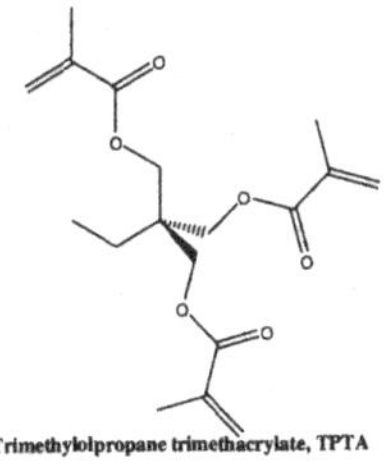

Trimethylolpropane trimethacrylate, TPTA
3-branch cross linker, to increase cross linking density

Figure 3 Examples of cross-linkers

Two commercial available cross-linking agents, the thermally curable binder ethoxylated bisphenol A dimethacrylate esters (EBAD) and the UV curable binder NOA83H, were investigated for incorporating hole transport material ADS132 into the polymer network to form the HTL. The HTL also acts as an electron blocking layer (EBL). First, the solvent resistibility of the spin-coated films from the mixture solution of ADS132 and cross-linking material were tested by washing the film with chloroform and toluene. In both cases using EBAD and NOA83H, the photoluminescence of the films before and after washing remained the same. No change of film thickness and morphology was found after washing the films with the solvents

Representative devices and their device characteristics are shown in Figures **4** and **5**. Sample A doesn't have a HTL/EBL, while sample B and sample C have ADS132 embedded in EBAD and NOA83H, respectively, as HTL material. The selected HTL material, ADS132 has a LUMO energy level that provides a barrier to effectively block electrons coming from the cathode. This electron blocking layer (EBL) drastically reduces the leakage current as evidenced by I-V characteristics of Fig 4B & C vs compared with Fig 4A, which tremendously improves external emission efficiency. The UV cured binder NOA83H (for ADS132GE) shows better performance than thermally cured binder EBAD in blocking leakage current (electron blocking) by 100X to 400X (compare Fig 4B & 4C). In addition to the effect of electron blocking from ADS 132, the low current may be also due to the the extremly high resistivity of NOA83H.

HTL/EBL that uses the NOA83H binder also has higher luminescent efficiency. For example, when operated at 4 volt, the EL peak intensity of NOA83H (5C) is 1/3 and 1/30 that of PFO only (5A) and EBAD (5B) respectively (**Fig. 5**), but using only 1/1000 and 1/400 the amount of current respectively (**Fig 4**). Therefore, the device with HTL in NOA83H has the highest external quantum efficiency compared to PFO only (without HTL/EBL) and HTL in EBAD device, which results from the large effect of the leakage current reduction. The EL spectrum distortion shown in device B with EBAD binder might be the result of deep penetration of PFO into the porous EBAD layer; Another possible explainnation is that the excitons are generated at different interface compared with PFO only devices os that the emission may come from ADS132. The different dielectric environment from the surrounding (EBAD and ADS132GE) could modify the energy structure of PFO. The absolute external quantum efficiencies of these devices at bias of 5V are listed in **Table 1**. As shown in Table 1, the external quantum efficiency for PFO only device is only 0.17%, while adding a layer of HTL of ADS132GE inside cross-linked EBAD and NOA83H polymer matriax increased the efficency to 2.94% and 8.71%, respectively. That's about 17 fold and 51 fold increasement on device efficiency for EBAD and NOA83H respectively. The attempt of making device with ITO/PEDOT/ADS132GE/PFO/Ba-Al failed due to the solublity of ADS132GE in most common solvents such as toluene and chloroform. The enhancemet of electroluminescnet efficiency of adding HTL is probably due to a more balanced electron and hole injection by introducing a HTL. This result demonstrates the effectiveness on HTL structural protection offered by the cross-linked NOA83H and the preservation of HTL electrical property that was designed for.

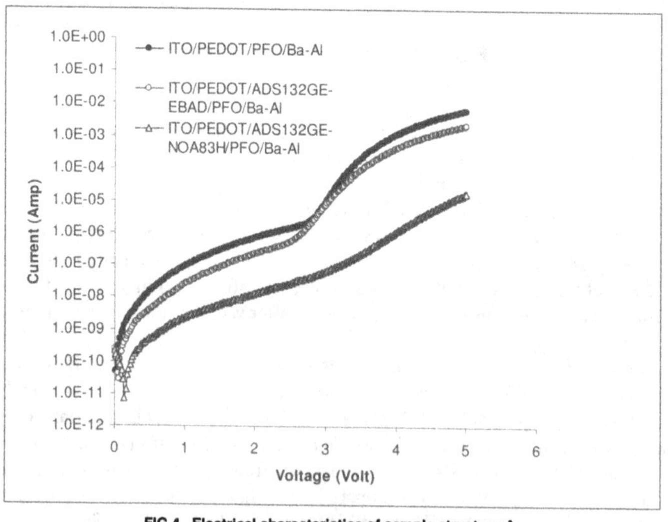

FIG 4. Electrical characteristics of sample structure A: ITO/PEDOT/PFO/Ba-Al, B: ITO/PEDOT/ADS132GE-EBAD/PFO/Ba-Al, and C: ITO/PEDOT/ADS132GE-NOA83H/PFO/Ba-Al.

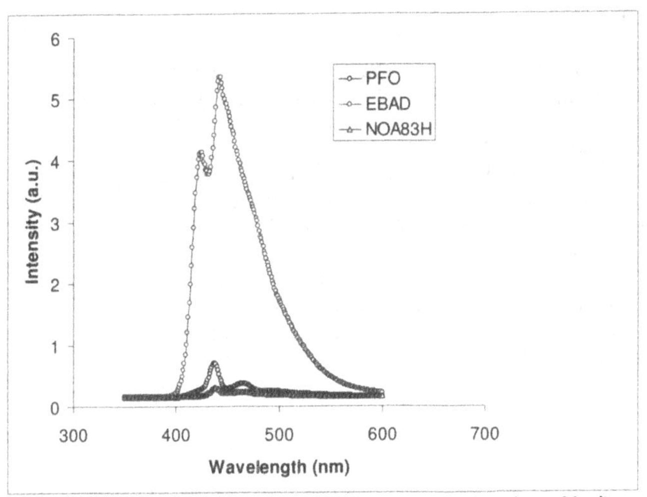

FIG 5 EL intensity and Spectrum of three devices A, B and C at bias voltage of 4 volts.

Table 1 The external quantum efficiency of three devices at bias of 5V

Entry	Device architecture	Extenal quantum efficiency
1	ITO/PEDOT/PFO/Ba-Al	0.17%
2	ITO/PEDOT/ADS132GE-EBAD/PFO/Ba-Al	2.94%
3	ITO/PEDOT/ADS132GE-NOA83H/PFO/Ba-Al	8.71%

It is noteworthy that we chose PFO as the emissive layer to prove the feasibility of our concept. Other small organic molecules or polymer light emitting materials, quantum dots (QD) and polymer/QD hybrid emitting devices could be easily used. We are currently developing the polymer/semiconducting nanocrystal hybrid pixel for improved life time, even better color purity and possibly power efficiency.

Conclusions

We have successfully fabricated multilayer functional electroluminescent pixels using a process of embedding HTL materials into cross-linkable agents. We have the fine-tuned formulation of cross-linkable monomers to produce smooth film layers, which shows good performance in PLED sample devices. The electroluminescence efficiency is enhanced by the more balanced electron and hole injection with this type of HTL. This result demonstrates the effectiveness of the HTL structural protection offered by the cross-linked NOA83H and the desired preservation of HTL electrical properties. This process will be very useful for solution – based multi-layer PLED device fabrications.

REFERENCES

1. J. G. C. Veinot and T. J. Marks, Acc. Chem. Res. **38**, 632 (2005).
2. H. Yan, B.J. Scott, Q. Huang and T. J. Marks, Adv. Mater. **16**, 1948 (2004).
3. H. Yan, P. Lee, N. R. Armstrong, A. Graham, G. A. Evmenenko, P. Dutta and T. J. Marks, J. Am. Chem. Soc. **127**, 3172 (2005).
4. S. Lee, Y.-Y. Lyu and S.- H. Lee, Synthetic Metals **156**, 1004 (2006).
5. J. Gui, Q. Huang, J. G.C. Veinot, H. Yan and T. J. Marks, Adv. Mater. **14**, 565 (2002).
6. B. Lim, J. T. Hwang, J. Y. Kim, J. Ghim, D. Vak, Y.-Y. Noh, S. H. Lee, K. Lee , A. J. Heeger and D.-Y. Kim, Org. Lett. 8, 4703 (2006).
7. J. Zhao, J. A. Bardecker, A.M. Munro, M. S. Liu, Y. Niu, I. –K. Ding, J. Luo, B. Chen, A. K.-Y. Jen and D. S. Ginger, Nano Letters **6**, 463 (2006).
8. E. Bacher, M. Bayerl, P. Rudati, N. Reckefuss, C.D. Muller, K. Meerholtz and O. Nuyken, Macromolecules **38**, 1640 (2005)

Processing, Sensing and Memory

Mater. Res. Soc. Symp. Proc. Vol. 1154 © 2009 Materials Research Society 1154-B11-05

Introduction of Innovative Dopant Concentration Profiles to Broaden the Recombination Zone of Phosphorescent OVPD-Processed Organic Light Emitting Diodes

M. Bösing[1], C. Zimmermann[1], F. Lindla[1], F. Jessen[1], P. van Gemmern[2], D. Bertram[2], N. Meyer[3], D. Keiper[3], M. Heuken[1,3], H. Kalisch[1], R. H. Jansen[1]
[1]Chair of Electromagnetic Theory, RWTH Aachen University, Kackertstr. 15-17, 52072 Aachen, Germany
[2]Philips Technologie GmbH, Philipsstr. 8, 52068 Aachen, Germany
[3]AIXTRON AG, Kaiserstr. 100, 52134 Herzogenrath, Germany

ABSTRACT

OLED with non-constant dopant concentration profiles have been processed by means of organic vapour phase deposition (OVPD) and were compared with regard to their luminous current efficiencies. Especially when driven at ultra-high luminance (>10,000 cd/A), OLED with a dopant concentration profile starting with a rather high dopant concentration on the anode side of the emissive layer showed improved luminous current efficiencies compared to their conventional counterparts.
To further investigate this effect, the width and location of the recombination zone have been simulated for all investigated concentration profiles by numerical solution of the semiconductor device equations using experimentally determined doping-dependent charge carrier mobilities. The obtained theoretical results are discussed with regard to the accomplished experiments.

INTRODUCTION

The introduction of phosphorescent emitters has established a basis for the development of OLED with impressive luminous efficiencies [1][2]. However, to exploit the full potential of phosphorescent light emission, it is crucial to develop device structures which lead to a relatively broad exciton recombination zone, in order to avoid triplet-triplet-annihilation (TTA) processes [3][4][5]. In many OLED, matrix and emitter contribute differently to the mobility of holes and electrons within the emissive layer, so the concentration of the dopant can be used to influence the position and shape of the recombination zone within the emissive layer. In contrast to vacuum thermal evaporation (VTE), OVPD intrinsically offers the opportunity to vary the concentration of the dopant during the deposition of the emissive layer, so that complex concentration profiles can be realized.
In this work, we use non-constant dopant concentration profiles, to locate and broaden the recombination zone of green phosphorescent OVPD-processed OLED in order to increase their luminous efficiency.

EXPERIMENT

To test the concept of non-constant dopant concentration profiles, two different OLED structures were employed: A rather simple three-layer structure (structure 1) consisting of a 40 nm emissive layer sandwiched between NPB (N,N'-diphenyl-N,N'-bis(1-naphthylphenyl)-1,1'-biphenyl-4,4'-diamine) and Alq_3 (tris-(8-hydroxyquinoline) aluminum) as hole and electron transport layer respectively, as well as a five-layer structure (structure 2) with a 10 nm emissive layer and additional exciton blocking layers (Fig. 1). Six OLED based on structure 1 and four OLED based on structure 2 with different constant and non-constant dopant concentration profiles were processed and compared with regard to their luminous current efficiency. The dopant concentration profiles of all OLED employing structure 1 are displayed in Figure 5 (together with simulation results of the associated recombination zone). The dopant concentration profiles of all OLED based on structure 2 are displayed in Figure 2 (HOMO energy levels are based on cyclic voltammetry (CV) data).

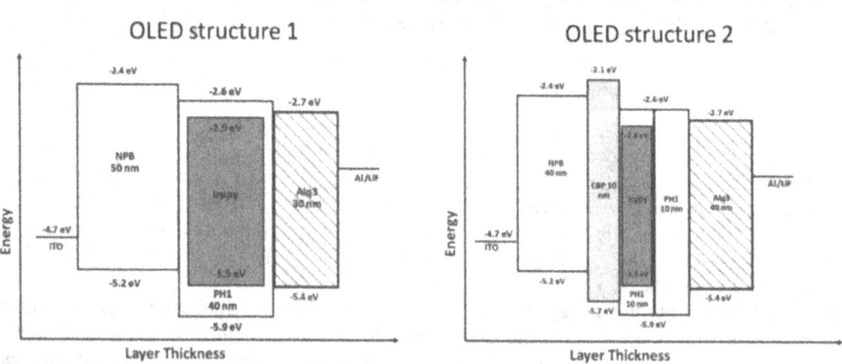

Figure 1: Energy diagram of green OLED

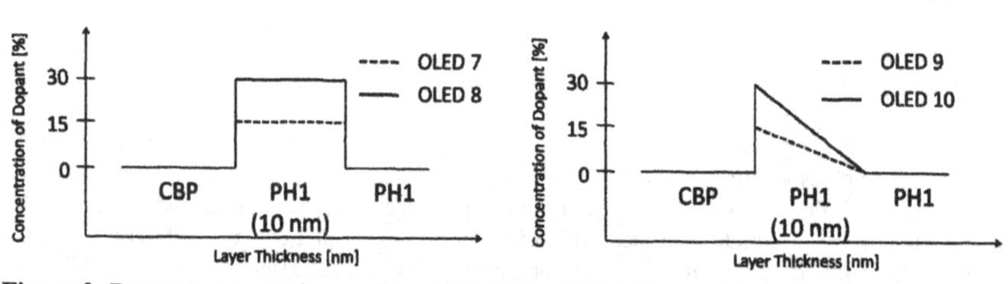

Figure 2: Dopant concentration profiles of OLED7 - OLED10

To successfully simulate the recombination of charge carriers within the emissive layer, the influence of the Irppy concentration on the hole and electron mobility must be known.

To investigate the influence of the Irppy concentration on the transport properties of charge carriers within the emissive layer, hole-only devices and electron-only devices with Irppy concentrations between 0% and 30% have been deposited on ITO-covered glass substrates. A 100 nm aluminum layer is used as a cathode. In the case of the electron-only devices, a 0.5 nm LiF layer is deposited prior to the aluminum cathode to improve electron injection. Since no light emission from the presumptive single carrier devices could be observed, it is guaranteed that the electron current in the hole-only devices as well as the hole current in the electron-only devices can be neglected within the regarded voltage range.

The hole-only devices consist of a 40 nm PH1:Irppy layer, sandwiched between two 20 nm NPB layers acting as hole injection layer and electron blocking layer, respectively (Fig. 3; left).
The electron-only devices consist of a 40 nm PH1:Irppy layer, sandwiched between a 20 nm PH1 hole blocking layer and a 20 nm Alq_3 electron injection layer (Fig. 3; right).

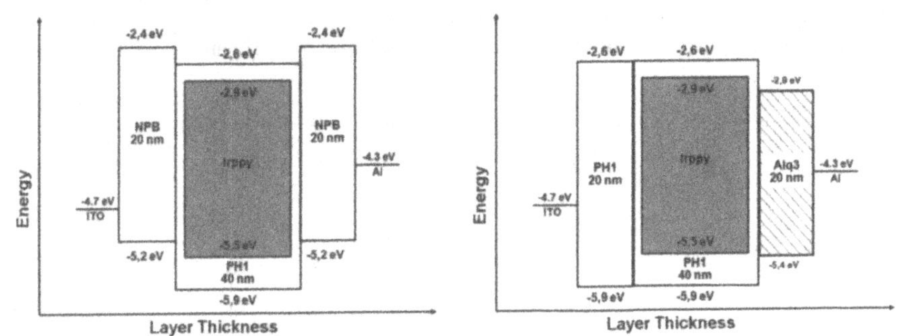

Figure 3: Energy diagram of hole-only devices (left) and electron only devices (right).

DISCUSSION

IV curves of all single carrier devices are displayed in Figure 4. For the hole-only devices, the current density is strongly correlated to the Irppy concentration. (Reducing the Irppy concentration from 30% to 4% results in a decrease of the current density by about five to six orders of magnitude). For the electron-only devices however, the current density is inversely correlated to the concentration of the dopant. Thus we suppose that within the emissive layer, the hole transport is mainly sustained by the dopant, whereas the electron transport is sustained by the host material PH1. Consequently, the dopant concentration is an eligible parameter to control the charge carrier recombination zone.

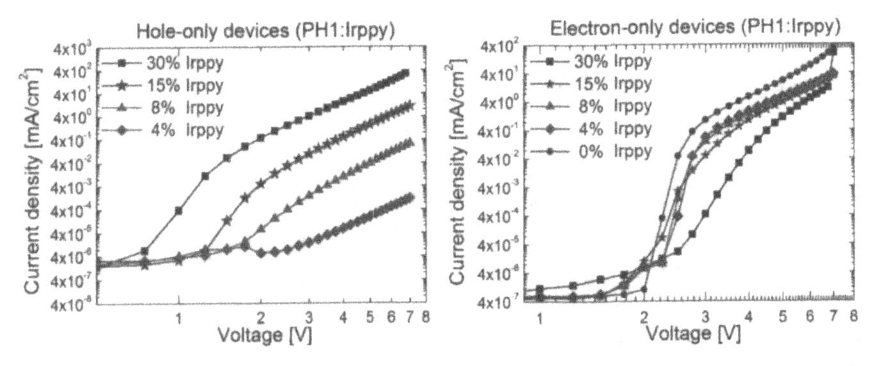

Figure 4: IV curves of single carrier devices with different Irppy concentrations

The recombination zones for the six devices based on structure 1 were simulated by solving the Poisson and the continuity equation. The dependency of the hole mobility on the dopant concentration was derived from IV curves of hole-only devices (Fig.4; left) assuming space charge limited current. Since the dopant concentration had little effect on the IV curves of the electron-only devices (Fig. 4; right), electron mobilities were assumed to be independent of dopant concentration. Energy barriers according to Fig. 1 and well-injecting electrodes were assumed. Diffusion currents were neglected. The results of the simulations are displayed in Figure 5.

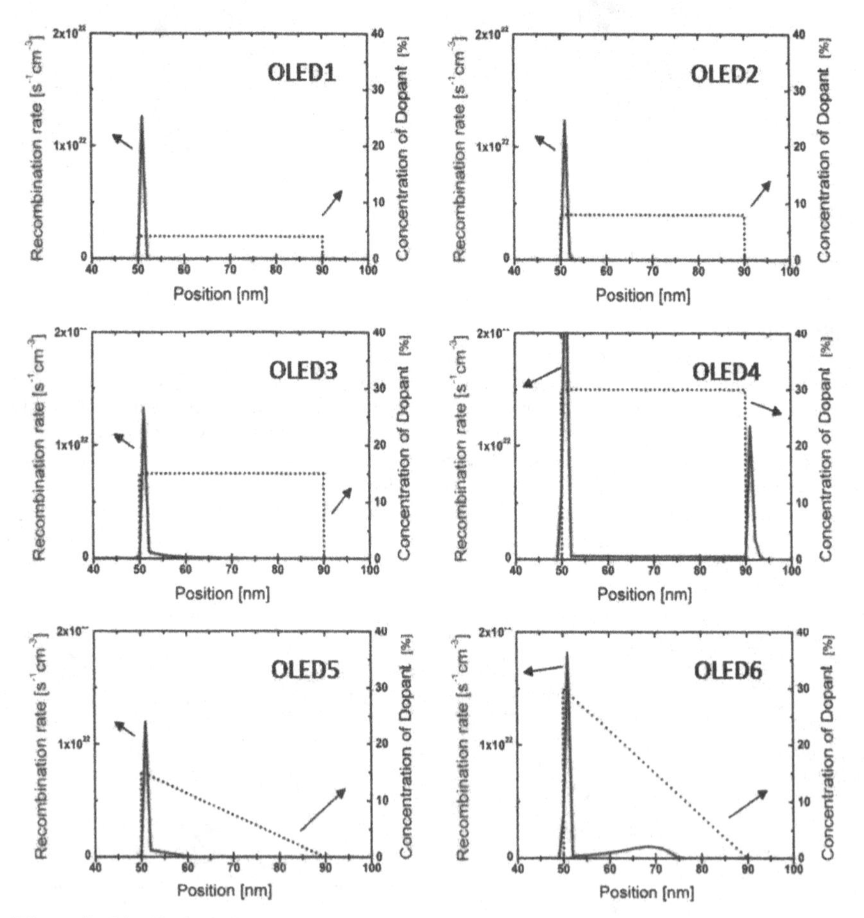

Figure 5: Distribution of recombination rate at a bias of 4 V according to simulation (red) and dopant concentration profiles (blue)

According to the simulation, OLED6 seems to be the most favourable of the six structures. Even though the recombination zone still displays a pronounced peak at the anode side of the emissive layer, about half of the recombination occurs in the middle of the emissive layer.
Luminous current efficiencies of all six devices employing structure 1 are displayed in Fig. 6 (left). As inferred from the simulation results, OLED6 shows the highest current efficiency of all devices employing structure 1. At a luminance of 27,000 cd/A, OLED6 shows a current efficiency of 17.2 cd/A which is 26% higher than the efficiency of the best device with a constant dopant concentration profile (OLED3: 13.6 cd/A).

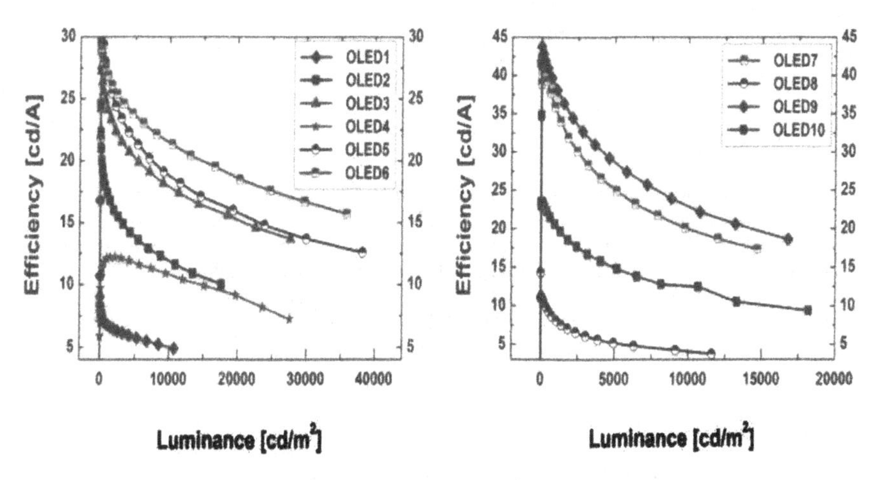

Figure 6: Luminous efficacy of OLED structure 1 (left) and OLED structure 2 (right) with different dopant concentration profiles

OLED1 and OLED2 both exhibit reasonably high maximum current efficiencies which decrease sharply at higher luminance. Since the simulated recombination rate of these two devices shows a very narrow peak at the anode side of the emissive layer (Fig. 5), we suppose that this roll off is due to TTA, caused by a high concentration of triplet excitons as well as a diffusion of triplet excitons from the emissive layer into the adjoining NPB layer.
In the case of OLED3 and OLED5 the simulated recombination zone is broadened compared to OLED1 and OLED2. Thus, the improved efficiency of these devices can be explained by a reduction of TTA.
According to the simulation, the charge carrier recombination in OLED4 occurs partly outside of the emissive layer and therefore explains the low efficiency of OLED4 even at low luminance.

Luminous current efficiencies of all four devices employing structure 2 are displayed in Fig. 6 (right). Again, the best result was obtained with a non-constant dopant profile, but in contrast to structure 1, a lower initial dopant concentration is favourable (OLED9).

Device	Current efficiency [cd/A] @ 1,000 cd/m^2	Current efficiency [cd/A] @ 10,000 cd/m^2	Current efficiency [cd/A] @ 27,000 cd/m^2
OLED3 (const.)	24.1 cd/A	17.8 cd/A	13.6 cd/A
OLED6	27.0 cd/A	21.5 cd/A	17.2 cd/A
OLED7 (const.)	35.7	19.9 cd/A	-
OLED9	38.5	23.0 cd/A	-

Table 1: Luminous efficacy of OLED structure 1 (top) and OLED structure 2 (bottom) with different dopant concentration profiles

CONCLUSIONS

Luminous efficiencies of both of the above employed OLED structures were improved by means of non constant dopant concentration profiles. The improvement is especially significant at a high luminance of more than 10,000 cd/m^2. Therefore this concept will gain relevance as future materials with improved lifetime allow the development of ultra-high brightness OLED.
According to the results of the simulation, the recombination zone of OLED6 (which was the most efficient OLED based on structure 1) still displays a pronounced peak at the anode side of the emissive layer and therefore leaves room for further improvements.

ACKNOWLEDGMENTS

The authors acknowledge the financial support by the German Ministry of Education and Science (BMBF) under grant no. 13N8669 (OPAL).

REFERENCES

1. D. Tanaka, H. Sasabe, Y.-J. Li, S.-J. Su, T. Takeda, and J. Kido. **Jpn. J. of Appl. Phys.** Vol. 46, No. 1, 2007, pp. L10–L12
2. F. Lindla et al. **MRS Symp. Proc.** 2009, in publishing
3. M. A. Baldo, C. Adachi, and S. R. Forrest. **Phys. Rev.** B 62, 10967 (2000)
4. R. G. Kepler, J. C. Caris, P. Avakian, and E. Abramson, **Phys. Rev. Lett.** 10, 400 (1963)
5. S. Reineke, K. Walzer, and K. Leo **Phys. Rev.** B 75, 125328 (2007)

AUTHOR INDEX

Printed in the United States
By Bookmasters